A CASE FOR CLIMATE ENGINEERING

A CASE FOR CLIMATE ENGINEERING

David Keith

A Boston Review Book

THE MIT PRESS Cambridge, Mass. London, England

MIT Press books may be purchased at special quantity discounts for business or sales promotional use. For information, please email special_sales@mitpress.mit.edu or write to Special Sales Department, The MIT Press, 55 Hayward Street, Cambridge, MA 02142.

This book was set in Adobe Garamond by *Boston Review* and was printed on recycled paper and bound in the United States of America.

Library of Congress Cataloging-in-Publication Data

Keith, David W.
A case for climate engineering / David Keith.
pages cm. — (Boston review books)
Includes bibliographic references.
ISBN 978-0-262-01982-8 (hardcover : alkaline paper)
1. Climate change mitigation. 2. Global warming—Prevention. 3. Environmental engineering. 4. Sulfuric acid—Environmental aspects. 5. Environmental geotechnology. 6. Enviornmental policy. I. Title.
QC903.K45 2013
551.68—dc23

2013027459

10 9 8 7 6 5 4 3 2 1

For Alex and Sarah,
my best hope for the future

CONTENTS

PREFACE

It is possible to cool the planet by injecting reflective particles of sulfuric acid into the upper atmosphere where they would scatter a tiny fraction of incoming sunlight back to space, creating a thin sunshade for the ground beneath. To say that it's "possible" understates the case: it is cheap and technically easy. The specialized aircraft and dispersal systems required to get started could be deployed in a few years for the price of a Hollywood blockbuster.

I don't advocate such a quick-and-dirty start to climate engineering, nor do I expect any such sudden action, but the underlying science is sound and the technological developments are real. This single technology could increase the productivity of ecosystems across the planet and stop global warming; it could increase crop yields, particularly those in the hottest and poorest parts of the world. It is hyperbolic but not inaccurate to call it a cheap tool that could green the world.

Solar geoengineering is a set of emerging technologies to manipulate the climate. These technologies could partially counteract climate change caused by the gradual accumulation of carbon dioxide. Deliberately adding one pollutant to temporarily counter another is a brutally ugly technical fix, yet that is the essence of the suggestion that sulfur be injected into the stratosphere to limit the damage caused by the carbon we've pumped into the air.

Solar geoengineering is an extraordinarily powerful tool. But it is also dangerous. It entails novel environmental risks. And, like climate change itself, its effects are unequal, so even if it makes many farmers better off, others will be worse off. It is so cheap that almost any nation could afford to alter the earth's climate, a fact that may accelerate the shifting balance of global power, raising security concerns that could, in the worst case, lead to war. If misused, geoengineering could drive extraordinarily rapid climate change, imperiling global food supply. In the long run, stable control of geoengineering may require new forms of global governance and may prove as disruptive to the political order of the 21st century as nuclear weapons were for the 20th.

Many people feel a visceral sense of repugnance on first hearing about geoengineering. For some, the repugnance crystallizes into moral outrage against the very idea that the topic is being discussed; for others, exposure to debate about geoengineering brings with it an appreciation of the hard choices at its roots and an understanding that there are credible arguments for and against. That intuitive revulsion strikes me as healthy; our gadget-obsessed culture is all too easily drawn to a shiny new tech fix. A narrow focus on a technology's power too easily blinds us to its risks.

It's tempting to wish climate change away by denying the science or by asserting that a quick shift to new clean energy sources provides an easy way out. But there is no magic bullet. We cannot make sound decisions by supposing the world is as we wish it were: the science of climate risk is solid, and the inertia of the carbon cycle combined with that of the world's economy mean that efforts to cut emissions can only moderate (but not reverse) climate change over this century.

As with the capacity to engineer our own genome, humanity is rapidly developing the capacity to engineer the planetary environment. Geoengineering's powerful potential demands a broad debate that must include not only credible arguments for and against such an intervention, but also, as with genetic engineering, an appreciation of the large questions it raises about nature and technology on a planetary scale.

I myself have concluded that it makes sense to move with deliberate haste towards deployment of geoengineering. You may well reach a different conclusion. My goal is simply to convince you that it's a hard choice.

* * *

In this book I attempt a synoptic view of solar geoengineering for the educated non-specialist who is willing to work their way through some complex arguments. I am not a disinterested bystander. Every author's story is shaped by their biases. The remainder of this preface discloses some of mine.

Wilderness has shaped my life. From weekend canoe trips to long solo ski expeditions in the high Arctic, I am fortunate to have spent about a year of my life traveling in the big wilds far from roads. My thinking was shaped by a family interest in environmental protection; my father played an early role in the science and regulation of DDT, and his brother helped lead the creation of birding as a social activity separate from scientific ornithology. I am an oddball environmentalist. A bit of a liberal redneck perhaps, as I have taken part in Earth First! actions and Christmas bird counts, yet I have a freezer filled with last fall's mule deer. I am also a tinkerer and a technophile. From lucking into a job at a top laser physics lab in high school, to teaching myself oxy-acetylene welding as I rebuilt the rusted frame of my first car, and then winning a prize for the best doctorate in experimental physics at M.I.T., I have always loved getting my hands dirty with hardware. Turning away from physics because it did not seem to engage real-world problems, I started to work in climate and energy before the end of graduate school. In 1989 I stumbled into geoengineering. I was drawn in by the lack of high-quality analysis of either the technology or the policy implications, a lack that seemed odd given the potential importance of geoengineering to the climate's future.

I have worked on this topic for most of my academic career. While my academic writing aims at objectivity and dispassion, here I venture educated guesses that go beyond what can be defended in an academic research paper. While I aim at objectivity I don't hide my personal views. I have done my best to draw a clear distinction my judgment about what the facts are from my personal, value-laden judgments about what we ought to do.

My passion for this topic is rooted in a concern that environmentalism has lost its way. The language of environmental advocacy has become increasingly technocratic. Calls for action rely almost exclusively on (seemingly) objective quantitative measures of cost and benefit that amount to a crude appeal to self-interest. We are urged to protect natural landscapes not because walking through them brings pleasure, but because of the ecosystem services they yield, services like oxygen and clean water. These arguments have merit, but I think they obscure much of what actually drives people's choices. If we are protecting a rainforest because it stores carbon or yields wonder drugs, then we should be happy to cut down the forest if some carbon storage machine or molecular biotech lab can better provide these services. If we are protecting a wetland for its ability to hold and purify water then we should be happy to replace it with a housing development if that development includes technologies for water storage and filtration that does these jobs better than the wetland. For me the utilitarian benefits of nature are a grossly insufficient measure of its value.

I also worry that debate about climate change has degenerated into trench warfare in which arguments are increasingly extreme with claims that climate science is a money-spinning fraud countered by claims that carbon emissions poses an immediate existential threat. As the extremes dig in, the battle has stagnated so that it now obscures many of the facts and moral values at the root of our choices. For me geoengineering matters both in its own right and because it encourages us rethink some of our root assumptions about the means and ends of climate policy.

A fuzzy love of nature is uncontroversial. We are saturated with soft-focus environmental imagery. Green exhortations have become white noise at the same time as new social technologies have accelerated the decline in people's day-to-day experience with the natural world. I suspect that Edward O. Wilson, the entomologist and writer, captured more than a grain of truth with his biophilia hypothesis, the idea that humans have an innate urge to affiliate with other forms of life. For me, the challenge is to craft an environmental ethic that recognizes non-utilitarian values in the natural world without asserting that these values trump all others and without making naïve claims of a sharp distinction between nature and civilization. Humans have shaped the natural world since long before the industrial era, before even the invention of agriculture. Stone Age hunters exterminated large animals in each new land they entered and the impact of these extinctions cascaded throughout the landscape. When humans arrived in Australia about fifty thousand years ago, to cite but one example, hunting pressure drove to extinction roughly 90% of all species weighing more than a typical human.[1]

Recognition of humanity's role in shaping landscapes that seem "natural" does not—for me—drain them of value or turn them into an artifact. Quite the opposite; in part, their value lies in the history of how they got the way they are, the co-evolution of nature, culture and technology. In *Rambunctious*

Garden, Emma Maris argued that environmentalists should abandon the obsessive defense of pristine nature in favor of an expanded environmental ethic that embraces the messy yet vibrant reality of landscapes shaped by human action.

I am simultaneously persuaded and repulsed by Maris' arguments, but I am convinced that we cannot come to sensible conclusions about the merits of geoengineering or about climate policy itself without engaging these hard questions at the interface of society and nature.

A note on terminology

Solar geoengineering, also known as solar radiation management (SRM) is the sole subject of this book. But the term *geoengineering* also describes the removal of carbon dioxide from the atmosphere, often called carbon dioxide removal (CDR). Before we dive further into solar geoengineering, it's worth a detour to explore the differences between these two classes of technology.

There is a host of ways to remove carbon dioxide from the air, from increasing the stock of carbon in soils and forests as farmers and foresters change their management practices to engineering methods that could speed up the geological weathering cycle in which the dissolution of alkali minerals into seawater pulls carbon from the atmosphere into the ocean. Technology for direct capture of carbon dioxide from the air might also play a role. Note that Carbon Engineering, a company I founded, is developing such technology, though our work aims to enable low-carbon fuels. (I address the conflicts of interest this raises in the next note.)

Solar geoengineering and carbon removal technologies are tools in our kit for managing climate risk, a toolkit that includes: consuming less (conservation), providing the same services with less inputs of energy (efficiency), supplying energy with less carbon (decarbonization), removing carbon after it's emitted (carbon removal), engineering climate change at a given level of greenhouse gases (solar geoengineering), and, finally, reducing the impacts of climate change by taking extra measures to adapt as it changes (adaptation).

A last word on climate policy jargon: anything that cuts emissions of greenhouse gases below what they would have been in some mythical business-as-usual world is called mitigation. Mitigation thus includes conservation, efficiency and decarbonization.

Solar geoengineering and carbon removal each provide a means to manage climate risk and each are sometimes called geoengineering, but they are only somewhat more related to each other than either of them is to other tools in the toolbox. First, they are wholly distinct with respect to the science and technology required to develop, test, and deploy them. And second, they distribute cost and risks very differently. Solar geoengineering has negligible deployment costs and entails benefits and risks that are regional to global in scale. Carbon removal technologies have costs that are generally similar to mitigation, in that it might cost about a hundred dollars to remove a ton of carbon dioxide using technologies that accelerated the weathering cycle and about the same amount of avoid a ton of emissions by building wind-turbines to replace fossil fueled electricity. Like mitigation they have local risks at the point of use that must be traded against global (but not local) benefits with essentially no corresponding global risks.

This divergence of costs and risks means that the challenges solar geoengineering and carbon removal raise for policy and governance are almost wholly different. Carbon removal is like mitigation in that it requires policy incentives to balance local cost and risk against a global benefit that accrues in the distant future (as we will see when we explore the inertia of the carbon cycle). Indeed, until humanity's net emissions are zero any carbon removal method has precisely the same effect on the climate as mitigation—a ton not emitted is the same as a ton emitted and recaptured.

Because solar geoengineering and carbon removal have little in common, we will have a better chance to craft sensible policy if we treat them separately.[2] For the remainder of this book I will use *geoengineering* to describe solar geoengineering only.

A final note about money and conflict of interest:

My work on solar geoengineering has been funded by academic research grants and by a personal grant from Bill Gates for whom I act as an occasional informal advisor on climate change and energy technologies. All my work on this this topic is academic with open publication and no patenting.

I also have my hands dirty running Carbon Engineering, a small start-up company that is developing technology for direct capture of carbon dioxide from the atmosphere. We hope this technology will make it cheaper to reduce carbon emissions from parts of the transportation infrastructure such as aircraft that are otherwise hard to decarbonize, and we see ourselves competing with other ways to accomplish this goal, such as biofuels.

I see a sharp distinction in the role of private enterprise in solar geoengineering and carbon removal. The development of solar geoengineering technologies should be as public and transparent as possible. The extraordinary global power of these technologies means that they cannot be effectively governed by the local rules appropriate for more conventional technology. I believe that private, for-profit development (and patenting) of the core technologies for solar geoengineering should be strongly discouraged.

Direct carbon capture from the atmosphere is very different. Succeed or fail, the technology we are developing in Carbon Engineering is a contained industrial process with local risks similar to other industrial energy or mineral processing technologies. Our job at Carbon Engineering is to develop a technology but not to decide how or if it's used. Because of my involvement I do not claim to speak as an academic about carbon capture from the air.[3] Governments ought to regulate it in the public interest, and it's wrong for corporations (who are not people!) to preempt that regulatory role.

1

Engineering the World's Sunshine

How might geoengineering work? Suppose the goal was to cut the rate of global warming in half starting in 2020 by putting sulfuric acid into the stratosphere. If combined with serious efforts to cut emissions, this is—in my opinion—a plausible scenario for managing the human and ecological risks of climate change in a world without politics. It is doable in the narrow technocratic sense that a well-managed program could likely be ready to start by 2020 if given a provisional go-ahead today and a total budget of roughly a billion dollars.

This scenario is a tool with which we may explore the possible. It is not a prediction or plan. There are a handful of plausible geoengineering technologies and a host of ways they might be used including some that are immensely destructive. As a newly emerging and divisive idea, geoengineering will add to the confusion of climate politics. Though I hope to persuade you that it's a good idea in the sense that it might be approved and swiftly implemented in some ideal global democracy, the chance of starting this scenario by 2020 seems slight. We lack the social tools to make sound collective decisions about planetary management. I will explore the politics that may shape real-world outcomes later in this book.

To be effective, the sulfuric acid needs to be in tiny watery droplets—about a thousand times smaller than the width of a human hair—and it needs to be put into the stratosphere about 20 kilometers above the earth's surface. Once there, the droplets will scatter sunlight back into space reducing the amount of sunlight reaching the ground. This slight shading effect tends to cool the planet, partially offsetting the warming effect of greenhouse gases such as carbon dioxide. Water droplets would do a fine job scattering sunlight—any cloud does this—but they just don't live long enough in the dry air of the stratosphere. The reason for using sulfuric acid is simply to keep the droplets from evaporating. Once formed the acid droplets remain in the stratosphere for about a year before falling into the lower atmosphere, so they must be continuously replenished.

As carbon dioxide and other greenhouse gases accumulate in the atmosphere—and accumulate they will until humanity cut its emissions to nearly zero—the amount of sulfur needed to offset their warming grows year by year. After the first year of operation, the project would need to inject about 25 thousand tons per year of sulfur in order to offset half of that year's growth in warming due to that years accumulation of greenhouse gases. The next year one would need to use 50 thousand tons to provide enough cooling to offset the half the warming from two years growth in the atmospheres stock of greenhouse gases. In 2030, after a decade of operation the injection rate would need to be 250 thousand tons per year.[4]

It is not technically difficult to get sulfates into the stratosphere. The hardware to do this does not exist today, but it is nevertheless fair to say that the capability exists today in the sense that the required aircraft and dispersal technology could be engineered and built within a few years by many aerospace companies or governments. The technical challenge is not the sulfur dispersal hardware but rather the development of the science and observing tools to monitor the effectiveness and side-effects of a sulfate geoengineering program, but here too there are many tools that could be applied quickly.

Injection of sulfates might be accomplished using Gulfstream business jets retrofitted with off-the-shelf low-bypass jet engines to allow them to fly at altitudes over sixty thousand feed along with the hardware required to generate and disperse the sulfuric acid. Only one or two aircraft would be needed to start the program, and after a decade it would take about ten aircraft to lift the required 250 thousand tons each year at an annual cost of about 700 million dollars. It would then make sense to convert to purpose-built aircraft with longer wings better suited to high-altitude flight; this change would cut costs roughly in half and might allow global distribution of sulfate from two airfields.[5]

In 2070, after a half century of operation, the program would need to be injecting a bit more than a million tons per year using a fleet of a hundred aircraft, though by that time it might make sense to have switched from sulfuric acid to an engineered particle with fewer environmental impacts. Sulfates in the stratosphere are certainly not the best possible technology, but under the slow ramp scenario describe here there would be ample time to develop alternatives before the problems with sulfates are likely to become too severe.

You may find this scenario intriguing or crazy, but it's hard to argue that it's technically infeasible. The necessary hardware could be ready by 2020 and even after a half century the direct cost of the program would be less than one percent of what we now spend on clean energy development.

How well would it work? What are the risks? These questions are a central focus of this book, so consider the following as a teaser. There is no reasonable doubt that sulfate aerosols could be used to cut the average rate of global warming in half for the next half-century. But carbon dioxide is not a simple thermostat. Accumulating carbon dioxide warms most of the world most of the time, but its effect is not uniform. Warming will, for example, be stronger in the polar regions than near the equator. Climate change is not just warming, rainfall will increase and become more intense, and some regions will dry while others become wetter. And carbon's footprint extends beyond climate, it will acidify the ocean and fertilize the growth of some plants, and here again the effects will be unequal.

The addition of sulfates to the stratosphere does nothing to stop ocean acidification, and cannot perfectly compensate for the multidimensional climate changes produced by greenhouse gases. But there is very strong evidence that it can substantially reduce many of the most important climatic changes and their associated risks. Geoengineering does not simply moderate the increase in global average temperature; it can substantially reduce changes in both precipitation and temperature on a region-by-region basis. It could be used to halt or even reverse the loss of Arctic sea ice.[6]

While one can say with reasonable confidence that geoengineering could reduce many of the large-scale physical aspects of climate change, less is known about its ability to reduce human impacts, the knock-on effects of changing climate that are of greatest concern. Only a few studies have examined its effectiveness in limiting climate's impacts on agriculture, for example, but these studies have found that geoengineering will likely reduce at least some of the crop losses that are a predicted result of climate change so that crop productivity in some of the hottest—and poorest—regions of the world would be higher with an appropriate amount of geoengineering than they would be with the same amount of greenhouse gasses but no geoengineering.

Crop losses, heat stress and flooding are the impacts of climate change that are likely to fall most harshly on the world poorest. The moderate amounts of geoengineering contemplated in this slow ramp scenario are likely to reduce each of these impacts over the next half century, and so it will benefit the poor and politically disadvantaged who are most vulnerable to rapid environmental change.

This potential for reducing climate risk is the reason I take geoengineering seriously. This potential is uncertain, but estimates of the impacts of climate change are themselves uncertain, and these two uncertainties are tightly linked because the science we depend on to predict the risks of accumulating carbon in the atmosphere is, in good measure, the same body of science that forms the basis for understanding the potential of geoengineering.

The risks are real too. Injecting sulfates into the stratosphere will likely increase the damage to the ozone layer and when the sulfate particles descend into the lower atmosphere they will contribute to air pollution. Any climate change will be bad for some regions and groups while it benefits others. This is true for both carbon-driven warming and sulfate-driven cooling. The best guess is that geoengineering will reduce the damage from climate change in most regions, but even in the best case geoengineering will make some regions worse off, perhaps by increasing drought. The risk is therefore that the ability of sulfates—or other solar geoengineering methods—to limit regional climate change may be less than models predict, so that the damages will be larger.

Our knowledge of the risks is not purely theoretical. Real-world experience gives confidence that those risks can be understood. To understand the risk of injecting a million tons per year of sulfur into the atmosphere, for example, we can study the 1991 eruption of Mt Pinatubo, which put eight million tons of sulfur into the stratosphere. And each year humans pump roughly fifty million tons of sulfur into the atmosphere as air pollution. This is not an argument that we should ignore the risk of putting one million tons per year of sulfur into the stratosphere for geoengineering, but it should give confidence that there is a strong empirical basis on which to assess these risks, and it is a reason to expect that risks will be comparatively small.

We are not leaping into the unknown with a totally new chemical. Nor are the risks irreversible as in, for example, releasing a species with novel engineered genes. Sulfates do not reproduce. If you stop injecting them they will disappear from the stratosphere in a few years.

Still you might well think that the risks are too large and the benefits too uncertain to justify a commitment to geoengineering. I agree. Doubly so because there is thin record of serious analysis of geoengineering, and too much of what is available has been done by a small group of scientists (I am one of them), a group dubbed the "geo-clique" by science writer Eli Kintisch in *Hack the Planet*. The danger of group-think is real.

Indeed were we faced with a one-time choice between making a total commitment to a geoengineering program to offset all warming and abandoning geoengineering forever, I would choose abandonment. But this is not the choice we face. Our choice is between the status quo—with almost no organized research on the subject—and commitment to a serious research program that will develop the capability to geoengineer, improve understanding of the technology's risks and benefits, and open up the research community to dilute the geo-clique. Given this choice, I choose research; and if the research supports geoengineering's early promises, I would then choose gradual deployment.

The risks and benefits of geoengineering are not simply an objective property of the technology, they depend on the way it's used. The scenario I propose departs from total commitment in two crucial respects. First, it supposes a slow ramping up of the sulfate burden, giving us a better chance to see problems and fix them or stop sulfate injection before they grow into disasters. Second, it would aim to offset only half the warming. This is important because, as we will see, if you use enough sulfate to cancel all the warming, you decrease global precipitation and increase the risk of various damaging side-effects. Using geoengineering to offset only half the warming tips the scales away from risks and towards benefits. Offsetting only half the warming also has the advantage of preserving a direct incentive to cut emissions, a factor I see as crucial to any sensible climate policy.

Critics will claim that I am overselling geoengineering, and their concerns will likely turn on three legitimate issues. The first is real disagreement about the science. Some might argue than in an effort to overcome what I see as overblown skepticism of geoengineering, I may overstate the benefits and understate the risks. The second reflects assumptions about how it will be used. Out of convenience or perhaps prejudice most of the climate model studies of geoengineering in effect simulate the total commitment case in which it is used to compensate for all warming, exaggerating the risks in comparison to the more restrained implementation that I explore here. The final concern regards misuse, including the idea that geoengineering would become a substitute for efforts to cut emissions. I share this fear. Indeed I see misuse of geoengineering in a chaotic multi-polar world as its most serious risk by far.

If geoengineering is our sole strategy—if we keep emitting carbon dioxide so that it keeps accumulating in the atmosphere—we would have to continually increase the amount of solar geoengineering with (eventually) disastrous consequences. But this fact simply means that we must (eventually) stop emitting greenhouse gases. It does not mean that one should prematurely abandon a potentially useful tool to limit its impacts just because it too has limits. I turn to the social implications of geoengineering later, but here I explore the basic risk-benefit trade-offs in a world in which it would not wrongly deployed to avoid cutting emissions.

Geoengineering complements emissions reductions. Cutting emissions reduces the long run risk by stopping the accumulation of carbon, while geoengineering—if it works as expected—will mask risks in the short run (in the slow moving world of carbon and climate, *short run* means the next half century). But geoengineering cannot eliminate the underlying risk that comes from humanity's rapid (in geological time) transfer of carbon from underground reservoirs to the atmosphere. It's hard to overstate the importance of geoengineering's ability to reduce risk for current generations as there are no other methods that can reduce these risks significantly in the next half century.

When used in concert with emissions cuts, geoengineering can reduce risk without any need to geoengineer forever. Suppose emissions were steadily reduced so that they hit zero in 2070 causing the climate to warm until the latter part of the century and then stay warm for many centuries thereafter. If geoengineering was slowly ramped up until 2070 and then ramped back down over the following half-century to stop in 2120, the net effect would be to spread the same amount of warming out over a the full century, reducing the rate of warming by half. Because many climate impacts are very sensitive to the rate of warming, this could provide a big benefit even though it does not change the total amount of warming over the century.

The idea that geoengineering could be temporary is contentious. Prominent critics such as geophysicist Ray Pierrehumbert have stated that once started geoengineering must continue forever, concluding that the whole idea is therefore "barking mad." Despite strong assertions, there do not seem to be physically-based argument to support the view that once started geoengineering cannot be stopped, particularly in the case in which it is ramped slowly up and then down.[7]

Until recently, controversy around geoengineering has largely suppressed research. Lack of research in turn sustains controversy because it allows exaggerated claims about benefits and risks to go unchallenged. Theory is ever a more fertile ground for extremism than the messy soil of reality. I contend that the potential upsides of geoengineering are large enough to justify an immediate commitment to a serious international research program that includes small-scale outdoor experiments. Research may show that these technologies will not work, yet the sooner we find this out the better. If research bears out the early promise of geoengineering then we have the option to begin gradual implementation if the political will—or necessity—is there.

2
Climate Risk

James Hansen, the world's most visible climate-scientist turned activist, has said that we must stop development of the Canadian tar sands, a growing source of energy-intensive oil production, or else, "It's game over for the planet."[8] I share Hansen's opposition to tar sands development, but I see no scientific basis to make such claims of imminent doom.

Predictions of catastrophe are more than rhetorical flourishes. They appeal to widely shared concerns that industrial civilization has lost its way, millennial fears that overconsumption is leading us to an environmental apocalypse that would be nature's ultimate act of retribution against a misguided species.

Our climate choices would be easy if we were truly facing an imminent existential threat. Emergencies entail extreme measures, a narrow focus on a single problem that may justify suspension of democratic due process. Imagine how effectively the world might collaborate if we discovered a massive asteroid inbound for a 2050 impact. But, this is not what we face. The claim that climate change threatens an imminent catastrophe is an attempt to play a trump card of (seemingly) objective science in order to avoid debate about the trade-offs at the heart of climate policy and about the role that values play in driving each of our personal judgments of the moral weight we accord to competing interests. But climate change is one of many problems, so there is no substitute for realistic assessment of our risks and open debate about the trade-offs between them, for we cannot avoid all risk or solve all problems.

If there is no imminent existential threat, why do we want to stop climate change? Because climate risks are serious. While it's true that change is the only constant in the long evolution of the earth's climate, industrial humanity has become a geological force driven change at unprecedented rates. Combustion of fossil fuels has raised carbon dioxide to levels not seen for millions of years. And the pace is accelerating; roughly half of all the carbon dioxide emitted in human history was emitted in the last quarter century. If we continue on our present course, within this century—within the lifetime of my children—human actions may well push carbon dioxide concentrations to levels not seen since the Eocene, the last period in which geoscientists are confident that carbon dioxide concentrations stood above 1,000 parts per million, about four times their pre-industrial value of about 270 parts per million. The Eocene was an era in the geologic record that ended about thirty-five million years ago, an era in which crocodilians walked the shores of Axel Heiberg Island in the present-day Canadian Arctic.

Global emissions of carbon dioxide now exceed thirty billion metric tons per year, an average of five tons per capita. An average American is responsible for about twenty tons, four times the global average. This is a big number, the weight of about ten cars for each person each year. The mass of carbon dioxide we dispose of in the atmosphere dwarfs by a factor of forty the mass of garbage we send to landfills. It now exceeds all human-driven material flows, including the gigantic movements of mine overburden. If carbon dioxide were garbage—a smelly mass that had to be pushed around with bulldozers—we would have dealt with this problem long ago.

The root of the climate threat is that humanity is moving carbon from deep geological reservoirs to the biosphere approximately a hundred times faster than the corresponding natural process in which carbon escapes from the Earth's crust at locations such as volcanic vents. While significant uncertainty remains about the climate's sensitivity to increasing carbon dioxide and about the attribution of recent warming to the historical increase in carbon dioxide, the unprecedented human acceleration of the carbon cycle is an undisputed fact.

We cannot accurately predict the results of our uncontrolled "experiment" on our planet, but it is certain that if we continue on our present course we are committing our children to climate changes that will be extraordinarily rapid compared to those humanity has experienced over the ten millennia since the invention of agriculture, when carbon dioxide concentrations stayed within about 5 percent of their preindustrial average. It is plausible, for example, that if we continue emitting carbon dioxide at the current rate, the sea level could rise more than five meters within a few centuries, enough to dramatically alter coastlines. Low-lying areas such as Florida would be inundated.

There is nothing wrong with the Eocene climate; there is no inherent reason we should prefer our crocodiles in the Florida Keys rather than on Axel Heiberg Island. The climate risks come from the rate of change, not because the current climate is some magic optimum for life. Our infrastructures, our crops, the very locations of our coastal cities have evolved for the current climate.

While social science cannot accurately predict the economic and social repercussions of rapid climate change, and while some sectors or regions will no doubt benefit, the losers will outnumber the winners. Analysis of climate impacts often focuses on the aggregate economic damages which are thought to run to a few percent of the world's economy, but this implicitly assume that the winners will compensate the losers. Maybe. Or perhaps the sociopolitical tensions arising from this rapid reshuffling of the deck will pose the largest risks.

Uncertainty + Inertia = Danger

Climate change is a particularly dangerous and difficult problem because of way it combines inertia and uncertainty. Climate responds to the accumulation of emissions in the atmosphere. And like the accumulation of water in a bathtub, the level does not drop quickly when one shuts off the tap. Like the bathtub, carbon is a stock and flow problem: atmospheric carbon is the stock, humanity's fossil fuel combustion is the dominant inflow and the slow removal of carbon by the ocean is the drain.

Warming in 2030, for example, is driven by historical carbon emissions up to 2030, so even if we started a massive program of emissions cuts today that was able to ramp emissions down to zero by 2030—an all but inconceivable industrial transformation—there would be surprisingly little reduction in warming by 2030 because much of the carbon that will be warming the world in 2030 has already been emitted during the last century. Carbon casts a long shadow onto the future: a thousand years after we stop pumping carbon into the air the warming will still be about half as large as it was on the day we stopped—assuming, of course, that we do nothing to engineer the climate.

Beyond the carbon cycle's physical inertia we must contend with economic inertia. We suffer from the persistent illusion that we can rapidly accomplish the deep structural change necessary to decarbonize our economy. The illusion is fed by the fact that as consumers we buy clothes, electronics, and cars—"capital stock" to an economist—that have very short lifetimes compared the capital stock that forms the very skeleton of our industrial economy: factories and the world-spanning distribution networks for materials, fuels and electricity.

Few of us see this infrastructure first hand; few of us work in power plants and refineries, so we don't appreciate in our guts how much money and material is invested in them nor do we know how long they live. Construction on Ford's River Rouge complex began in 1917, and the plant is still a central node Ford's vehicle production infrastructure. The fleet of coal and nuclear power plants that supplies most US electricity is just a few years younger than the average American citizen.

This is not an excuse for inaction. Quite the opposite, the inertia of our energy system is a reason we should have started its transformation to low carbon alternatives decades ago when scientists first discovered the extent of carbon-climate risk.

Ending carbon emissions means ending conventional use of fossil fuels—the energy source that drives our civilization. We know how to make energy without carbon by using sources such as wind, nuclear, solar, or fossil fuels with CO_2 capture and geological storage. But switching to emissions-free energy is not like a consumer-driven switch between commodities. It's not Coke versus Pepsi or even SUVs versus small cars. Cutting emissions to zero means replacing a big chunk of the heavy infrastructure on which our society rests.

If we stopped making gasoline cars today and switched to electric vehicles, it would take about a decade before most cars on the road were electric. And, a switch to electrics only enables deep cuts in emissions if we decarbonize electricity supply. The combined inertia of the carbon cycle and the human economy is so great that if we began dramatic efforts to decarbonize our energy system today, we would not see most the benefits of reduced climate change until late this century.

Then there is uncertainty. Warming is *proportional* to emissions over this century in the sense that if we double our emissions we will roughly double the warming. The catch is that we can't accurately predict the amount of warming we will get for a given amount of emissions, the "climate sensitivity." Suppose we emitted a thousand billion tons of carbon this century. Humanity's carbon emissions are now ten billion tons per year and rising, so a thousand billion tons is the amount we would get if we stopped the growth and kept emissions at their current rate over the century. With luck the resulting warming might be below 1 degree centigrade—still almost twice the warming the world experienced over the 20th century—but with roughly equal odds the dice might roll against us and global warming might be above 4 degrees—a huge climate shift equivalent to half the warming between the depths of the ice age and today's climate. The most likely outcome is somewhat above the 2 degree mark that is widely discussed as an upper bound of acceptable warming.

Of course, this is an optimistic case because without dramatic efforts to restrain carbon emissions we will not hold this century's total emissions to the zero-growth thousand-billion-ton scenario.

Worse still, the link between carbon emissions and the effects of climate change, such as crop losses or sea level rise, is even more uncertain. Uncertainty cascades from carbon emissions to risk. Consider sea level rise: first there is the uncertainty connecting emissions of carbon dioxide (that we control) to the concentration of carbon dioxide accumulating in the atmosphere.[9] Next we have the uncertainty in the climate sensitivity discussed above. Then we have uncertainty in predicting climate change over the big ice sheets and in the oceans that warm them, and finally uncertainty in the response of the Greenland and Antarctic ice sheets to that warming.

The low-probability but high-consequence "tails" of the probability distribution curve—the small chance that the dice rolls against us—account of much of overall climate risk. Moreover, we can't estimate the uncertainty very well: we don't know how much we don't know.

Together, uncertainty and inertia make the climate problem dangerous: if we are unlucky and climate change is particularly severe, by the time we realize the climate dice have rolled against us it will be too late to resolve our problems with a deathbed conversion to clean energy.

Cutting Emissions

There is good news. Governments of the world signed the Framework Convention on Climate Change in 1992 and ratified it within two years. The convention binds its signatories to stabilize "greenhouse gas concentrations in the atmosphere at a level that would prevent dangerous anthropogenic interference with the climate system." Spending on clean energy is skyrocketing: over the last few years the world has spent about a quarter trillion dollars per year on clean energy—mostly wind and solar power. This is about one a third of one percent of global economic output. It is very roughly in line with the spending that economic models suggest is needed to put us on track to meeting the convention's objective of stabilizing greenhouse gas concentrations by cutting emissions to (nearly) zero over half a century.

Yet despite all the treaty-making and windmill-building, emissions are rising fast. Global carbon emissions have grown at about 3% per year since 2000. There was a brief downturn for the 2008 recession, but emissions are now growing about as fast in this century as they did in the in the decade before the 1972 oil crisis, an event widely seen as turning point away from exponential growth towards energy efficiency and alternative energies. Rather than slowing, the growth of emissions has accelerated since the Framework Convention was signed. Emissions have consistently been at the upper end of the formal projections of emissions that are used to predict future climate change.

Recall stock and flow: emissions change the climate when carbon dioxide and other gases accumulate in the atmosphere. The concentrations of accumulated carbon dioxide in the atmosphere will hit 400 parts per million in the next year. While a round number means nothing to nature, to pass that mark with the concentration growing at more than 2 parts per million each year (and accelerating) is a powerful symbol of collective failure. Bill McKibben is doing a brilliant job organizing climate activism around 350.org, but the chance of stopping before 500 or of holding the rise in global temperatures under the 2 degree centigrade target seem remote.

Much of the recent emissions growth has come from China, but even in Europe there are disturbing signs that the required emission cuts may not be achieved. While emissions in the European Union have declined slightly since 2000, the rate of decline is not nearly fast enough. Moreover, the European Emissions Trading System, the capstone of Europe's plan to cutting emissions, is crumbling; and despite huge spending on renewables European coal consumption grew by 2% during 2012.[10] Coal is without a doubt the worst offender for both climate and human health. Closing coal-fired power plants is the first step in any rational plan to tackle climate change. There is no better measure of failure than the fact coal consumption is up in Germany, a leader in green policies that likely has world's highest per capita spending on clean energy.

If you could take an omniscient look across the world's energy system to assess the pace of actions—not rhetoric or symbols—to cut emissions you would likely be forced to conclude that such actions were all but negligible.

Why is there such disconnect between efforts and outcome? Why has the spending on clean energy produced such meager results? Either the cost of cutting emissions is much higher than analysts' estimates of what's needed or the money is getting grossly misspent. Carbon emissions are so large that deep cuts can only be realized by actions that are cost-effective and scalable. Much of the money for clean energy has been spent on projects with very high abatement cost—dollars per ton of carbon dioxide—such as solar photovoltaic on rooftops in Germany, corn ethanol, or Chinese wind farms built in locations with poor wind resources and inadequate connection to the electric grid. I think of projects such as solar panels on fancy homes as "green bling," they may serve as powerful symbols of green commitment but do very little help the climate.

The bitter truth is that the world's efforts to cut emissions have (with a few exceptions) amounted to a phony war of bold exhortation and symbolic action. It's tempting to assert emissions cuts are impossible and that we must look to alternatives like geoengineering. This is doubly wrong. First, solar geoengineering may reduce risks in the short term but it cannot get us out of the long-term need to cut emissions. Second, to assert that emissions cannot be cut is to take human agency—and responsibility—out of the picture as if emissions were coming from some species other than our own.

It is possible to cut emissions to near zero in half a century maintaining a high—and increasing—standard of living. Emission are not cut by rhetoric, but by improving energy efficiency and replacing high-emissions energy infrastructure such as coal-fired power plants with low-emission energy like solar, wind or nuclear power. Exploring the means to make such a deep emissions cut is beyond the scope of this book, suffice it to say that it might require a level of spending similar to the 5% that the U.S. spends on the military, and the money would have to be spent with a tough-minded discipline to find low-cost options. A realistic plan might include strong consumer efficiency standards; a substantial economy-wide carbon tax matched by reductions in other taxes, particularly taxes for the poor, who feel energy cost the most; an aggressive commitment to research and new clean energy technologies; and finally, the elimination of targeted subsides for boutique technologies that do not provide cost-effective emission reductions. No one can claim to know the best mix of clean energy technologies, but my hunch is that it will be hard to make rapid emissions cuts without some non-variable, high-intensity, low-carbon energy sources such as nuclear power or fossil fuels with carbon capture and storage. Finally, moving fast enough would require quick decision making of a kind that now seems out of reach of many democratic governments in the West.

But it could be done. While 5% of GDP is real money, it's about half of what the United States spent on its military in the 1960s and less than a third of what it now spends on healthcare. To claim that rapid emission cuts are impossible is to shirk responsibility for our actions.

3
Science

Science cannot resolve political or ethical questions about solar geoengineering, but neither can you develop sound judgment without a firm grasp on a few central facts. Sound judgment requires facts and ethics. Don't trust my ethical judgment on geoengineering just because I happen to know a lot of facts; and likewise, don't trust assertions about facts from someone just because you share their ethical stance.

The essential questions of fact regarding solar geoengineering are easy to pose: How effectively can it counteract climate change? And what are the risks? What are the unknowns? On the question of efficacy, we know that climatic change produced by scattering some sunlight is not the same as the change produced by carbon dioxide and that fact will limit any solar geoengineering technology from fully counteracting climate change. Regarding environmental risk, we know that any method used to scatter sunlight will have side-effects, such as the loss of ozone caused by sulfate aerosol in the stratosphere. This chapter explores efficacy and environmental risk while questions about the ethics and politics of geoengineering will occupy the last two chapters of the book.

Political conflict too often morphs into proxy battles about what constitutes objective fact. Nowhere is this more evident than in the battle over climate policy, where legitimate differences about the role of government and hard trade-offs between the welfare of present and future generations are played out as shrill debates about the "hockey stick" temperature record. This is similar to the debate about geoengineering, where proponents minimize risks and opponents exaggerate them.

Efficacy: How effectively could we reduce climate risks if we could manage sunlight?

Any discussion of climate change rests on a distinction—often implicit—between climate forcing and climate response. Climate *forcings* are external factors that influence climate. A change in the intensity of sunlight arriving at the top of the atmosphere is called "solar forcing." Nothing the Earth does can influence the Sun's intensity, so a change in the solar constant is an external driver—a cause—of any resulting climate change.

The Earth absorbs sunlight and radiates the energy back to space as infrared heat. Nature always balances her energy budget, so if the Earth absorbs a bit more sunlight then temperatures must increase in order to bring the system back into balance by radiating away a bit more infrared heat to balance the extra heat adsorbed from the sun. The warming is a response to the solar forcing.

Carbon dioxide has an insulating effect making it harder for the Earth to radiate infrared heat, so when carbon dioxide levels rise the temperature must rise a bit to allow the planet to balance its energy budget. This is the essence of global warming: a response to climate forcing by carbon dioxide.[11]

Using this forcing/response framework, we can say that geoengineering is the intentional manipulation of climate forcings with the goal of counteracting undesired climate change. We will get to the technology later, but for now let's say that the most plausible method is, as I mentioned earlier, to add reflective aerosols such as sulfates to the stratosphere, where they reflect some sunlight back to space.

If our climate problems were due to the sun getting brighter we might use aerosol geoengineering to achieve a near-perfect fix for the resulting warming because the climate forcing from aerosols could—by design—be almost precisely equal and opposite to the climate forcing from increased sunlight. After all, the essence of solar geoengineering is to reflect away some sunlight.

Climate change from carbon dioxide cannot be so accurately offset because the radiative forcing by carbon dioxide is qualitatively different from radiative forcing by geoengineering. The positive radiative forcing from increased carbon dioxide, for example, acts during both day and night, whereas the negative forcing from reduced sunlight acts only in daylight. If one warms the climate up by adding carbon dioxide, one can cool it down to the original temperature by selecting the appropriate amount of geoengineering, but the resulting climate will have warmer nights and cooler days than the original unperturbed climate.

Day-to-night temperature differences are a minor issue for geoengineering, but they illustrate the physics behind the fact that reducing sunlight cannot perfectly restore a climate warmed by carbon dioxide. The same line of argument applies to the seasonal cycle because carbon dioxide's warming is felt about equally in summer and winter whereas cooling from geoengineering is strongest in summer when solar heating is most intense.

Given the sharp differences in the patterns of radiative forcing from carbon dioxide and geoengineering one might expect that geoengineering would do a lousy job countering climate change. The first climate model experiment to test the effectiveness of geoengineering grew out of an argument at a meeting in Aspen in 1998. Lowell Wood was extolling the virtues of geoengineering. Several of us took the bait, most prominently Ken Caldeira, who heckled Lowell from the back of the room, saying, in essence, "Don't you know any atmospheric science? The radiative forcings are very different so the climate compensation will be lousy."[12]

To Caldeira's lasting credit, he went back to his lab and tested his assumptions by running a climate model experiment, anticipating that he would show how poorly geoengineering worked and so prove Wood wrong. Instead, Caldeira found the reverse: geoengineering did a surprisingly accurate job in compensating for the climate change caused by increased carbon dioxide.[13]

Fourteen years after Caldeira's first model simulation, most of the major climate models have been used to simulate the effectiveness of geoengineering.

Caldeira's first simulation simply adjusted the model's solar constant to simulate a uniform solar dimming. Recent simulations are far more sophisticated. They include such things as state-of-the-art treatment of the physics of sulfate aerosols, tests of other geoengineering methods such as the whitening of clouds with sea-salt aerosol, and use "ensembles" comprised of thousands of individual climate models, each programmed with slightly different representation of atmospheric physics, as a tool to assess uncertainty.

Despite encouraging results, the essential fact remains: no plausible method of geoengineering produces a radiative forcing that is the exact opposite of the forcing from carbon dioxide. We could use geoengineering to maintain the planet's average surface temperature, or even to bring it back down to its pre-industrial value, but the resulting climate would not be exactly the same as the pre-industrial climate.

If we can't restore the pre-industrial climate with geoengineering, how far can we go in reducing the worst impacts of climate change? Climate impacts are local. No one feels the global-average climate. Just as temperature and precipitation respond differently to geoengineering, so too does the climate in different regions of the globe. Yet surprisingly little progress has been made in understanding the effectiveness of geoengineering in limiting regional climate risks such as extreme weather or loss of agricultural productivity.

Debate about the effectiveness of geoengineering too often bounces between advocates who tout the global average response and critics who identify regional effects where geoengineering makes things worse. Both arguments are valid, but they obscure the fact that the amount of geoengineered radiative forcing—and thus its effects—is a human choice.

Carbon dioxide warms the world, which increases precipitation. It also has a separate direct effect of suppressing precipitation. So while precipitation levels rise with levels of carbon dioxide, the net increase is less than it would have been if the world had been warmed by some other means, such an increase in the sun's intensity.

Suppose the world is warmed by carbon dioxide and then cooled by geoengineering until the average temperature is restored to its pre-industrial level. The resulting climate will have less precipitation than the original. This fact is often spun into claims that geoengineering will cause a decrease in precipitation and possible drought. This is wrong and misleading: it ignores carbon dioxide's role in suppressing precipitation; and it obscures the choice of how much geoengineering to use. It's true that if geoengineering is used to restore global average surface temperature to pre-industrial values, precipitation will decrease. But it's equally true that if a bit less geoengineering is used so that it's just enough to restore precipitation then temperature will increase. These views are two sides of the same fact: more geoengineered radiative forcing is needed to restore temperature than is needed to restore precipitation.

The false impression that geoengineering *will* reduce precipitation is fueled by fact that most climate model studies have chosen the amount of geoengineering so as to restore average temperatures to their pre-industrial value after a doubling of carbon dioxide. This may be convenient choice for running climate models output, but it's a crazy prescription for actual geoengineering. Many climate impacts arise from the rate of change: warm five degrees over a thousand years and almost all human activities and most ecosystems will have no trouble adapting. But warm the same five degrees over half a century and there will be big trouble. Sudden application of enough geoengineering to take us back to pre-industrial value (we have warmed about 0.8 centigrade above that) would itself have big impacts.

If it makes sense to use geoengineering at all, I think it will be in combination with cutting emissions and with a goal to reduce—but not eliminate—the rate of temperature rise. This strategy need not reduce average precipitation.

Just as it's possible to adjust the amount of geoengineering to restore pre-industrial temperature or precipitation, but not both, so too one can adjust it to minimize the change in climate in one region at a time, but not in all regions as once. Juan Moreno-Cruz and Kate Rickie, two of my graduate students, worked together to explore the inequality of geoengineering's impacts on a region-by-region basis. Our goal was to demonstrate how unequal effects might create geopolitical tensions by exposing the sharp trade-offs between the amount of geoengineering that would be best for one region as opposed to another.

Our results surprised me: geoengineering does a much better job on a region-by-region basis than I expected. We ran the model with increasing greenhouse gases to simulate the climate of 2030 and analyzed the changes in each of 22 standard regions that widely used for analysis of climate impacts, comparing each region's climate with that of the 1990s. As expected, the model forecasts dramatic changes with most regions getting more precipitation in the 2030s and about a quarter getting less. When we added geoengineering—and chose the amount sulfate aerosol to minimize changes in precipitation—we were able to reduce the average change by 87 percent. The average region had simulated 2030's precipitation within 17 percent of 1990's levels.[14] This result strongly contradicts the widely held view that geoengineering only works for temperature, or for global average quantities.

One caveat: these results are no better than the climate models they depend on. I have no confidence in precise numbers like 87 percent. But we can say that based on the same models used to predict climate change from greenhouse gases, geoengineering seems to be able to substantially reduce climatic change—and presumably climate impacts—on both globally and on regional basis, despite the inherently imperfect compensation that arises because climate forcings from greenhouse gases and geoengineering are intrinsically different.

Nonetheless, the possibility of different regional effects raises valid concerns about equity. In my view these concerns arise more sharply from questions about the legitimacy of decision-making in the use of solar geoengineering, rather than the inherent limits of the technique itself. Unfortunately, discussion about geoengineering itself has crystallized in a debate about the Asian Monsoon, the seasonal rains that come when warm air over India rises, pulling moist ocean air inland. Concern was first raised by scientist Alan Robock, who said that sulfate injection into the stratosphere "would disrupt the Asian and African summer monsoons, reducing precipitation to the food supply for billions of people."[15] Robock, an expert in volcanoes and the way eruptions alter climate by putting sulfate aerosol into the stratosphere, has become perhaps the most visible scientific critic of geoengineering.

Pundits have also weighed in on the threat of geoengineering to the Asian monsoon. More half of all Google hits for "geoengineering" now include "monsoon,"[16] and left-leaning pundits such as Arun Gupta argue that a technocratic elite (he cites me by name) threaten the lives of billions for profit: "Many scientists fear that pumping sulfates into the atmosphere may cause Asia's monsoons to fail, putting more than a billion people at risk of starvation."[17]

Indeed, used recklessly, geoengineering could threaten billions with starvation. However, Gupta's narrative ignores *all* studies to date (including Robock's) which suggest that the appropriate use of geoengineering could reduce climate risks to Asian agriculture.

The rhetoric gets a number of things wrong. First, most models of climate change show harmful precipitation *increases* in the Asian monsoon region (think floods and mudslides). If geoengineering can slow or stop this increase in precipitation, then it is delivering a benefit, not a harm. Again, the claim that geoengineering will reduce precipitation is based on an assumption that geoengineering will be used to stop the rise in temperatures. This is not a fact about geoengineering, but rather an assumption about how much it will be used.

Second, a reduction in precipitation need not lead to drought. The amount of surface water found in soils or as runoff in rivers depends on the balance between precipitation and evaporation. Real world impacts depend on soil moisture and runoff as well as precipitation, so looking at precipitation alone tends to exaggerate the impacts. In the global average the two must always be equal, since all the water that evaporates into the air must eventually return as precipitation to close the hydrological cycle. A climate with increased carbon dioxide that has had temperatures held to pre-industrial levels by geoengineering has less precipitation and less evaporation, but it may or may not have more drought. One must do the analysis before making that claim.

Third, critics overlook a related fact. Extreme climate events, such as droughts or floods, depend on the overall strength of the hydrological cycle. As the world warms, the hydrological cycle intensifies, causing an increase in extreme events out of proportion to the simple increase in average temperature and precipitation. This is one of the most important factors that drive climate impact in a warming world.[18]

By weakening the hydrological cycle, geoengineering may therefore be more effective at reducing climate impacts than one might expect by looking only at the changes in averages.

Finally, models run to date suggest that if used appropriately geoengineering could substantially increase food supply by reducing heat-stress during the early growing season in Asia and Africa—an effect that is one of the most important causes of crop loss as the world warms. In the only study yet published that used a state-of-the-art model of agricultural productivity, Ken Caldeira and colleagues found that solar geoengineering could increase yields of rice and wheat in India by 15–25 percent.[19]

Is worry about the Asian monsoon misplaced? No. If used foolishly, geoengineering *could* disrupt the Asian and African summer monsoons and threaten the food supply for billions of people. Robock was right to raise the issue. There is a lot we still don't know. Despite the host of computer model runs testing climate model's response to solar geoengineering, the research community is just now beginning to use these simulations to look at actual climate impacts on agriculture, ecosystems, and human health.

It's vital that debate about geoengineering shift from the physics of climate models to focus on human and environmental impacts. The risks of geoengineering need urgent attention, along with the benefits. But the current debate reveals the extent to which people fit geoengineering into convenient narratives while ignoring facts that don't fit their theory.

Signal and noise

Suppose we tried geoengineering for a decade, increasing the radiative forcing fast enough to offset half the warming from accumulated greenhouse gases. Would we know how well it worked, or how badly it failed? The short answer is no. Even after a decade of manipulating the climate at unprecedented scale we might learn remarkably little about the effectiveness of geoengineering.

In the language of physics, the "signal" of our experiment might not rise above the "noise" of natural variability.

Weather varies day-to-day. Climate is the average weather or, as an old saying goes, "climate is what you expect and weather is what you get." The climate varies decade-to-decade. Some of this variability is caused by variation in radiative forcing—such as the occasional volcano or a fluctuation in the sun's intensity—but some is random fluctuation that we call noise.

When the signal is smaller than the noise, it's hard to measure the signal even with good data. One might have a theory that some change in tax law will increase the value of certain stocks yet be unable to prove or disprove the theory in the face of the stock market's random noise.

The difficulty of finding signal buried in noise explains why, despite centuries of pumping carbon dioxide into the air, we still do not know precisely how it is changing the climate.

The problem is no mere technocratic detail, for it may shape the global politics of geoengineering. Suppose a coalition of countries begins geoengineering and then China is hit by unprecedented drought. Chinese officials would be very tempted to blame the coalition and seek compensation despite the difficulty of proving causation.

Confusion about signal and noise infects debates about near-term testing. Many argue that geoengineering cannot be meaningfully tested short of full-scale implementation.[20] While we can never be certain of the efficacy and risks of geoengineering from models alone, this claim is doubly wrong. First, we can learn a lot from testing it on a smaller scale if we concentrate on responses that have a large signal or little background noise as would be the case we look for changes in stratospheric aerosols and chemistry following injection of particles and second, even it were tested at "full scale" we will still not resolve all our uncertainties.

Aerosol geoengineering: how might we manage sunlight?

If we wanted to engineer large-scale radiative forcing at low cost with minimal side-effects, how would we do it? The most plausible near-term method is to increase amount of sulfuric acid aerosol in the stratosphere.[21]

An aerosol is simply a suspension of fine particles in air. A puff of household dust is an aerosol. So too is the spray of oil droplets injected into the cylinder of a diesel engine and the nanoparticles of ash that is left over when each oil drop burns. That ash adds to the burden aerosol in urban air, likely contributing to my asthma, which I counter with an aerosol of steroid sprayed into my lungs. The steroid aerosol is highly imperfect technical fix for the pollution aerosol, but both aerosols and imperfect technical fixes are ubiquitous.

Aerosol particles scatter light. That's why we see clouds but can't see the water vapor out of which they condense. It turns out that the amount of light scattered for each kilogram of aerosol is greatest if the aerosols are a few tenths of a micron across, about the size of transistors in your computer's CPU and about a thousand times smaller than a raindrop. Small aerosols fall out of the atmosphere much more slowly than big ones. Saying that an aerosol is a "suspension" is just a fancy way to say that the particles have not yet settled out. A cannonball released in the stratosphere would hit the ground in about two minutes, but pare it down to a tenth of a micron and you may wait more than two years for it to fall to earth.

The *stratosphere* gets its name because it is highly stratified, so that while air mixes horizontally it does not readily mix upwards or downwards. It is distinguished from the *troposphere*, or "turning sphere," the lower atmosphere in which we live, which is rapidly turned over as warm air rises and cold air sinks to take its place. An aerosol particle can remain in the stratosphere for years, whereas the same particle might last only days in the troposphere before being captured in a raindrop and brought to earth.

The stratosphere begins about 7 to 15 kilometers over our heads. It's highest in the tropics, and lowest near the poles and during winter. Commercial passenger jets cruise at an altitude of about 10 kilometers, so while they don't get into the stratosphere in the tropics one spends much of a winter flight between Europe and North America in the stratosphere.

Aerosols have very roughly the same ability to scatter sunlight back to space, wherever they are in the atmosphere; it takes about the same amount of aerosol in the troposphere or the stratosphere to produce a given radiative forcing. But, since aerosol lifetimes are about a hundred times shorter in the troposphere, one must add new aerosol a hundred times faster than would be required to maintain the same quantity of aerosol in the stratosphere. This is why stratospheric aerosol is preferable to tropospheric aerosol. To achieve the same radiative forcing, stratospheric aerosol would not only cost about a hundredfold less, and any side effects, such as acid rain or air pollution, would also be about a hundredfold smaller.

The immense leverage provided by stratospheric aerosols is evident in the ratio of carbon to sulfur. Only a few tons of sulfur in the stratosphere is needed to offset the radiative forcing of a million tons of carbon in the atmosphere. Consider this: Radiative forcing is measured in watts per square meter (W/m^2). Since the beginning of the industrial revolution, emissions of carbon dioxide have increased the atmospheric burden of carbon by about 240 billion tons. That added carbon produces a radiative forcing of about 1.7 W/m^2—about intensity with which a 60 W ceiling bulb illuminates the floor beneath it. A radiative forcing of –0.85 W/m^2, an amount sufficient to counterbalance half of the current carbon dioxide forcing, would require injecting about only one million tons of sulfur into the stratosphere each year to maintain the required amount of sulfate aerosol.

This near million-to-one leverage is at the root of both the risk and promise of stratospheric aerosol geoengineering; it is the underlying reason why it is such a powerful and frightening tool.

Risk

The best understood risk of sulfate geoengineering is ozone loss. The stratospheric ozone layer is the planet's natural protection against the sun's ultraviolet radiation, which would otherwise be strong enough to sterilize life on land.[22] (This "good" high-altitude ozone is the same chemical as the "bad" low-altitude ozone that forms in urban areas when sunlight cooks smog.)

Chlorine acts as a catalyst to destroy stratospheric ozone. Naturally occurring chlorine levels in the stratosphere are negligible. Essentially all the chlorine comes from the breakdown of chlorofluorocarbons (CFCs), a man-made chemical. The history of CFCs is itself a warning about technical fixes. Before the 1930s refrigerators used ammonia, which is sufficiently toxic that people died when refrigerators leaked coolant. Chlorofluorocarbons emerged as wonder chemicals, because they were inert and nontoxic. They were soon adopted for a range of uses beyond refrigeration, from propellants for spray cans to military fire extinguishers. But their very inertness caused a new problem. With no natural breakdown process to remove them from the lower atmosphere, CFCs gradually make their way into the stratosphere, where strong UV light breaks them up, releasing chlorine. That chlorine destroys ozone by providing a fast way for the extra oxygen atoms in ozone to link up, converting them back into oxygen.[23]

Aerosols interfere with ozone chemistry through a confusing maze of chemical pathways so that the addition of aerosol may either increase or decrease ozone concentrations. The primary pathway by which sulfuric acid droplets from geoengineering would reduce ozone is by reducing the amount of nitrogen oxides (NOx) in the stratosphere. NOx binds chlorine into a chemical called chlorine nitrate which does not catalyze ozone destruction. But sulfates facilitate the reactions that turn NOx back into non-reactive reservoirs of nitrogen. When aerosols reduce NOx they increase the fraction of chlorine that is actively destroying ozone.

The impact of geoengineering aerosols depends not only on what geonengineering aerosols we add to the stratosphere, but also on how much chlorine is there when we do. Chlorine loading is decreasing following the 1989 implementation of the Montréal Protocol and related treaties that phased out use of CFCs. The risk of ozone loss therefore depends on when we might use geoengineering; if we use it late in the century when CFC levels will be low, there will be little effect; if we use it sooner the effect could be significant. Used at very large scale by 2045, geoengineering would cut ozone by 10% near the poles, with smaller losses at mid-latitudes. But even with geoengineering, the ozone concentration in 2045 would be significantly higher (better) than today because stratospheric chlorine loading is expected to decline by 30 percent between now and 2045.[24]

There are other risks. The roughly 50 million tons per year of sulfur pollution we presently dump into the lower atmosphere kills about one million people per year through asthma, heart disease, and lung cancer.[25] Air pollution deaths per ton of sulfur used for geoengineering will likely be substantially less, since it would be distributed world-wide (air pollution from industries is concentrated near population centers). But, if we begin putting a million tons of sulfur into the stratosphere each year, it will probably contribute to thousands of air pollution deaths a year. Geoengineering would presumably save more lives than it takes by reducing climate risks that kill by heat stress, floods or famine. But this ratio does not settle the moral concerns as different people will die and the actions that cause deaths are organized in fundamentally different ways: deaths from air pollution that are caused by a myriad sources from cars to power plants whereas fatalities from geoengineering would be attributable to single (presumably) centrally organized program.

Sulfate aerosol is the devil we know. Because of the impact of sulfate aerosol pollution and of volcanos there is a solid body of science to inform our understanding of the risks of sulfate aerosol geoengineering. Sulfates pose additional risks, such as acid rain, beyond the ozone and health risks discussed above, but the risks studied to date appear relatively small when compared to the risks of rapid climate change. New risks will be associated with new methods of engineering radiative forcing, such as the possibility of using engineered nanoparticles, which I will discuss in the next chapter. The largest concern is not the risks we know but rather a sensible fear of the unknown-unknowns that may surprise us.

4

Technology and Design

Arguments for research on geoengineering often appeal to the ethic of scientific freedom. Discussion about its risks and efficacy are similarly couched in language that assumes that research will uncover the risks and efficacy as if they were facts of nature. This framing deeply misconceives the task at hand. In making science a passive discoverer-of-facts it buries the active role of the technology's developers. Are we trying to protect the arctic or enable the poorest and most vulnerable people to limit the damage they suffer from a changing climate? Do we want to actively engineer climate to maximize global crop productivity? The geoengineering-as-science framing stunts debate about the appropriate role of public involvement in forging the tools to manipulate climate.

Science does not discover tools to manipulate climate any more than science discovers the iPod. The development of both is an engineering task that starts with a conception of the problem to be solved and builds a new technology out of a storehouse of scientific knowledge and preexisting technologies.[26] Engineering may lead to new science, for the interaction between the two pursuits is richly bidirectional; but, while engineering and science are too intertwined to allow a clean demarcation, there is nevertheless an essential difference: engineering is about design and science is about understanding.

Design requires a designer. Every designer starts with some human need they aim to satisfy, and their conception of that need in turn drives their design. The designer may be a lone architect sketching a family house or a team of aerospace engineers designing a new guidance package for a shoulder launched anti-aircraft missile, but however misconceived it may be, the designer's conception of the problem to be solved plays a huge role in shaping the resulting technology.

Science is different. It is about understanding not design. And that understanding is measured in only one currency: the accuracy of its predictions. Richard Feynman once wisely defined science as "what we have learned about how not to fool ourselves." There is no universal method, just a bag of tricks for making better predictions and avoiding self-delusion.

Some of the more heated debate around geoengineering has implicitly assumed that we would be faced will an all-or-nothing decision to turn it "on" at full strength to manage some undefined "climate emergency." We need to break free from the too common assumption that geoengineering will progress linearly from research to deployment. Technologies develop in messy, multiply-connected loops. We should expect surprise; we may start with sulfate aerosols to limit temperature extremes yet end up with different technology and goals.

To understand the potential and risks of geoengineering one needs some insight into the choices that its designers will face and some specific ideas about how it might be used. To that end I offer a specific scenario for deployment. For the moment, imagine that this highly idealized plan will be managed by some apolitical agency staffed with the best and brightest from around the world, who act without bias or secrecy under the aegis of some legitimate international agreement. Later I will move beyond this naïve fantasy to explore how geoengineering might interact with the rough-and-tumble of real-world politics.

A scenario for deployment

Phase 1: Theory and laboratory work. The first step is simply to get serious about applying the science, social science, and technology we have at hand to understand the efficacy and risks of solar geoengineering. For the physical sciences, this means exercising the current suite of atmospheric models to understand what they say about such topics as the dynamics of aerosols in the stratosphere, the impact on ozone, and the climate change that results from radiative forcing by stratospheric aerosols. Beyond atmospheric science, we would apply the toolbox—such as historical analysis of the way crop yields or infectious diseases are influenced by climate—that has been developed to understand the impacts of climate change. One could then explore how the imperfect reduction in climate change afforded by geoengineering may reduce (or increase!) climate impacts from agriculture and glaciers to water supplies, biodiversity, and economic inequality.

Laboratory work might focus on understanding chemical reactions that seem to be particularly important to predicting the impact of aerosols on stratospheric chemistry and on developing tools to generate and monitor aerosols in the stratosphere.

Phase 2: Experiments in the atmosphere. The first atmospheric experiments should focus on understanding the process by which aerosols are produced and through which they may disrupt the chemistry of the stratosphere. Many of the key atmospheric chemical processes have a daily cycle, so observations of an artificial aerosol cloud over just a day or two could provide a powerful test of our understanding. Such experiments are small-scale tests of the physical processes that are built into large-scale models. Because the goal is to test processes, not the atmosphere's large-scale response to forcing, the amount of material needed for such experiments would be miniscule compared to the amount needed to alter the climate in measurable ways. Experiments that my collaborators and I are now contemplating would use less than a hundred kilograms of aerosol material—less than one ten-millionth of what we would need to add every year to make a readily measurable impact on climate. The experiments would be performed using the same scientific instruments and research platforms (balloons and aircraft) that have been used for decades to study the causes of ozone loss in the stratosphere.

How can experiments build understanding of risk without imposing the same risks? By studying component processes, as when stress tests on a single turbine blade enables engineers to predict engine failure without putting an airplane full of people a risk. But no amount of piece-by-piece testing can predict failures that emerge from unanticipated interaction of aircraft components or the pilot's decisions. So, too, no amount of process experiments can eliminate the chance of unexpected problems with geoengineering.

Some of my colleagues argue that the growing power of scientific models means that we need not go outside. I disagree. Computer models are no substitute for the real world. Environmental science is full of surprises that were only uncovered by careful observation. Theoretical predications can amaze: in 1974 Mario Molina and Sherwood Rowland predicted the impacts of chlorofluorocarbons via a novel chemical pathway, before any direct observations of ozone loss (a feat for which they shared a Nobel Prize). But theory has limitations. The ozone hole,[27] for example, was discovered, by a British team using painstaking ground-based observations from Antarctica. NASA satellites might have discovered the effect earlier but analysts were blinded by theory; they threw away the data showing very low ozone levels in the Antarctic spring because their data analysis software assumed that the readings must be instrumental error.

Beyond the chance of discovering the unexpected, outdoor experiments will, I anticipate, provide a better platform than mere theory on which to anchor debates about governance of geoengineering.

The work of Phase 1 would not stop when experiments began; rather it would accelerate to interpret the results of the Phase 2 experiments and assimilate them into improved predictions of geoengineering's efficacy and risks.

Phase 3: Minimal deployment. If, and only if, results from the first two phases warrant, the next step would be deployment at the smallest scale at which a response can be detected. The goal would be to find unexpected problems before they become big enough to cause damage. In a rational world, one would never try even minimal deployment unless results of the first two phases suggested that benefits of geoengineering outweigh side effects. Deployment crosses a threshold beyond science that demands some form of legitimate governance.

The question of how big minimal deployment needs to be is a question of signal and noise that the tools of Phase 1 would investigate. As a thought experiment, though, suppose one injects aerosols into the northern hemisphere stratosphere for about a year at an intensity sufficient to produce a cooling over land of a fraction of a degree centigrade. This would not quite be detectable above the random year-to-year temperature variation, but other effects, including changes in stratospheric temperature, the intensity and character of solar radiation, the surface energy balance of ice sheets and some ecosystem effects might be detected with careful monitoring.[28]

Again, the work of the first two phases would not end with Phase 3, as new small-scale experiments or laboratory work might be needed to understand the results of a minimal deployment.

Phase 4: Gradual deployment. If, after the first three phases, geoengineering still looks beneficial, one might begin gradual deployment. The goal would be to reduce climate risk, but not by suddenly "turning on" a geoengineered radiative forcing at the strength necessary to counteract all of human radiative forcing. That would cool the planet back towards the pre-industrial climate—a potential disaster. Most impacts of climate change, such as those on agriculture, depend more strongly on the rate of climate change than on the absolute amount of change, so a sudden, large scale-up would be worse than nothing.

Moreover, some unexpected bad side effect might only be detectable as the amount of geoengineering increases beyond an unknown threshold. So to control the rate of change and to maximize the chance of detecting problems early, gradually ramping up the radiative forcing from geoengineering makes the most sense.

Consider this specific scenario: suppose we ramp up the geoengineered radiative forcing at a rate sufficient to counterbalance half the increase in all other human radiative forcings. On a global-average basis, this would roughly cut the rate of climate change by half. Using sulfate aerosols to counter radiative forcing, one would need to increase the amount of sulfur added from zero to about 250 thousand tons per year over a decade.

To put the risks of this intervention in perspective: if we offset half of the current increase in radiative forcing of climate change for the next half century, the annual input of sulfur into the stratosphere in 50 years would be about 1.3 million tons per year. That's less than one-fifth the quantity of sulfur spewed into the stratosphere by the Mount Pinatubo eruption.

How quickly could all this happen? The kind of scientific research efforts described in Phase 1 have already begun, albeit in a slow and uncoordinated fashion. In a world without politics, I could imagine quickly ramping up these activities to produce a large-scale, loosely coordinated international effort within five years, at which point the first atmospheric experiments might begin. A deployment that would have measurable large-scale effects on the atmosphere (Phase 3) would not begin until about a decade from now, and it would take a minimum of half a decade of such experiments before we might intelligently begin a gradual deployment. That takes us to 2025, and, more realistically, 2035 with missteps and surprises.

The slow pace means that if one wants the ability to use geoengineering to protect us from some "unpleasant surprises in the greenhouse," as Wally Broecker, a tough freethinker and one of intellectual giants of earth science, put it in a 1987 *Nature* article, then we need to begin serious research now.[29] Indeed, we should have started long ago.

Artificial volcanoes

Most analysis of geoengineering has assumed that we would inject sulfur dioxide into the stratosphere, imitating the mechanism by which volcanoes alter climate. Sulfur dioxide is, however, a gas even in the chilly lower stratosphere where temperatures can get below –80 C, and a gas will not condense to form aerosol droplets. To make an aerosol, the sulfur dioxide must first oxidize to sulfuric acid in a natural process that takes about a month. Once created, the sulfuric acid molecules "want" to condense to form liquid sulfuric acid. If the molecules form near an existing droplet they will condense onto it, but if they cannot find a droplet quickly enough then sulfuric acid molecules will eventually find each other and begin a new droplet.

When a volcano injects sulfur dioxide into a "clean" stratosphere with few existing droplets, the result is that many new droplets produce an aerosol layer in which most of the sulfuric acid creates particles that are a few tenths of a micron across—a size at which they are very effective in scattering sunlight.

You would not achieve the same favorable aerosol size distribution if you pumped sulfur dioxide continuously into the stratosphere. Most of the added sulfur would end up finding existing droplets, increasing their size rather than making new ones. Size matters.

Larger aerosol particles fall faster, so they spend less time in the stratosphere and so scatter less sunlight over their lifetime. Moreover, the physics of light refraction in small particles dictates that as droplets grow larger than about two-tenths of a micron they become progressively less efficient at scattering sunlight. Taken together, these two effects mean that an aerosol cloud of large, micron-sized particles exerts dramatically less radiative forcing (cooling) over its lifetime than does a cloud in which the same amount of sulfur was distributed among smaller particles.

One possible solution to this "big droplet" problem is to release sulfuric acid vapor (rather than sulfur dioxide) from an aircraft. A plume of vapor released from an aircraft rapidly mixes with the surrounding air, but in the first seconds of mixing the concentration is so high that the vapor condenses into new small droplets even in the presence of background aerosol that would be present with continuous geoengineering. This "direct aerosol" sulfuric acid method may be significantly more effective than releasing sulfur dioxide because it limits the tendency of the new vapor to join existing droplets: in our simulations of this method less than 4 million tons of sulfur per year was enough to offset the radiative forcing from humanity's entire current greenhouse gas burden, more than twice as much sulfur is required with the sulfur dioxide method.[30] This matters, since less sulfur should mean less environmental risk.

These ideas are not new. In the 1970s Mikhail Budyko, an eminent Russian climate scientist, considered using sulfur dioxide to counter climate change driven by humanity's carbon emissions, an idea he dubbed "artificial volcanoes." Yet the "big droplet" problem was not discovered until 2009, when a research group in Zurich demonstrated it in computer simulations. Within a year I helped bring together scientists in Halifax, Boston, and Zurich to work on the direct injection of sulfuric acid vapor as a potential work-around.

All the science necessary to understand the big droplet problem and our proposed work-around has been known since Budyko's efforts. The fact that no one began to analyze the details of sulfate aerosol geoengineering until the last few years illustrates how money and intellectual fads shape the research frontier. For decades a de-facto taboo against serious work on geoengineering discouraged quantitative work; little was done. Paul Crutzen's 2006 paper[31] arguing for geoengineering research broke that taboo, not because of any new ideas, but because Crutzen, a Nobel Laureate, was credible and because his work on ozone chemistry—assumed to be the major risk of geoengineering—made it impossible to claim that he was ignorant of the risks.

The taboo had grown from a well-intentioned fear that talking about geoengineering would weaken the political drive to cut emissions. It was strong: several scientists wrote to the late Steve Schneider, then as editor of *Climatic Change*, begging him to reject Crutzen's article. Instead, Schneider, who had himself been thinking about geoengineering since the seventies and whose support helped spur me to work on the topic, asked several of us to write critical commentaries to accompany Crutzen's piece.

Only then did researchers begin to roll up their sleeves and get to work. Much of the progress in science comes not from breakthroughs but rather from workaday application of what we know to new domains. Our paper on the direct aerosol method was no breakthrough, but rather a pedestrian application of current knowledge to a new problem.

This is an important lesson. If we begin a serious research program, geoengineering will quickly shed its (illusory) simplicity. New problems will emerge and new work-arounds will be proposed. A crop of recent scientific papers finding new problems with the sulfate aerosols method suggests that this is already happening. But funding is still minimal because government research managers are rightfully cautious that such a small group of researchers dominate the research frontier. This is unhealthy, as diversity and competition are necessary to avoid groupthink. A sustained and diverse research program that reveals the messy complexity of geoengineering would, I believe, foster a healthier public debate and make it harder to defend the extreme positions that geoengineering is either a silver bullet or a reckless distraction.

Technology and cost to reach the stratosphere

How hard would it be to move one million tons of sulfur per year to the stratosphere? What would it cost? Popular writing is filled with the whizbang aspects of geoengineering hardware. Stories in media outlets ranging from *Popular Science* to the BBC debate the merits of naval guns, giant hoses suspended by balloon, dirigibles, and airplanes.

The excessive focus on deployment hardware arises from a healthy instinct: readers want a physical picture of what might be done and journalists find it hard to paint a simple graphic picture of stratospheric chemistry or climate response. In effect the whizbang treatment, however, distracts audiences from the hard questions of risk and trade-off that are largely decoupled from the particular choice of hardware. That said, I describe some of the likely deployment technologies here because I want to make clear that the technology is not the primary issue. Deployment is neither hard nor expensive.

First, we must understand how high one needs to get material for stratospheric geoengineering, since altitude plays a big role in determining the cost and difficulty of delivery technologies. While the stratosphere sometimes extends down below 10 kilometers (32,000 feet), where it is easily reached by passenger jets, air in this part of the stratosphere is rapidly mixed back into the lower atmosphere. Aerosol injected here would have a lifetime measured in months rather than years. This might be useful for the kinds of short-term tests described in Phase 2, but would not likely be sensible for large-scale aerosol geoengineering.[32]

The top of the stratosphere is about 50 kilometers (164,000 feet), far from reach of practical aircraft, but one need not go nearly so high. The natural flow of air in the stratosphere rises upwards from tropical latitudes and sinks downwards over the poles, so, if one can inject aerosol a bit above the bottom of the stratosphere in the tropics, it would be carried upwards and then towards the poles, resulting in a long lifetime and a relatively even distribution of aerosol throughout the stratosphere. Computer simulations confirm this intuition and suggest that injection in the tropics at altitudes a bit over 20 kilometers (65 thousand feet) would be adequate.

One could in principle use existing aircraft such as jet fighters, but modern business jets are more efficient and much cheaper. A stock Gulfstream G650, a top-of-the-line business jet, cruises at altitudes up to 50,000 feet. If a G650 were retrofitted with a low-bypass military engine such as the Pratt & Whitney F100, it could lift a payload of 13 tons to 60,000 feet, an altitude that would likely be adequate for the minimal deployment described in Phase 3. A fleet of just twenty aircraft acquired within a few years at a cost of $1.5 billion should enable sufficient radiative forcing to produce large-scale climatic effects that are just barely detectable.[33]

The requisite deployment technology does not exist as ready-to-go hardware today, but it could be supplied by any number of vendors using what the aerospace industry calls commercial off-the-shelf technology. We could build the deployment hardware far more quickly than we likely could develop the rest of the science, engineering, and governance required to begin deployment of geoengineering. In this sense that one can say that the technology exists today.

This is not an argument for or against immediate deployment. It is simply a statement of capability.

If, in a full deployment phase, millions of tons of material needed to be transported to the stratosphere on an ongoing basis, it would not make sense to use existing aircraft. Instead, one would build a small fleet of custom aircraft using conventional aircraft construction techniques and modified aircraft engine designs. The resulting aircraft would be similar in size to the business jet but would have dramatically longer wings, resembling those on the ER-2/U-2 that a played such a big role in stratospheric science (and geopolitics from the Cuban Missile Crisis to Iraq). Including investments in operations as well as the new aircraft, the cost to move materials to 75,000 feet is about 1 dollar per kilogram.

Another popular idea is the possibility of pumping material into the stratosphere using a long hose suspended by a stratospheric balloon. With sufficient effort it could likely be achieved, and it could be cheaper than aircraft. Hoses suffer from the disadvantage, however, that one cannot so easily spread particles around horizontally as will be necessary for some of the most obvious kind of stratospheric aerosols.[34]

The sole benefit of the stratospheric hose—if it can be made to work—is to make geoengineering still cheaper. Yet, cost is the one problem we do not have with this technology. Indeed, an early focus on making it cheaper sends precisely the wrong message. The UK government made an ill-considered decision in 2010 when it chose to fund a dramatic outdoor test of stratospheric hose technology as part of the Stratospheric Particle Injection for Climate Engineering (SPICE) project. The full SPICE project has many fine researchers doing worthwhile projects, but in 2011, public debate became quite properly focused on the hose experiment. In the most visible public fight about geoengineering, the ETC group, a leading critic of geoengineering, quite correctly dubbed the experiment the "Trojan Hose." This experience points to the urgent need for researchers to get more disciplined in tying experiments to clear socially-relevant goals, such as understanding the risk and efficacy of geoengineering.

Finally, it's certainly possible to use naval guns or rockets, but the cost to get to an altitude of 25 kilometers would be much greater than the cost for aircraft, so unless some new form of geoengineering aerosol is invented that requires delivery at much higher altitudes, it's unnecessary to consider such systems further.

Is it expensive? Taking the estimate of one dollar per kilogram delivered to 75,000 feet and assuming one million tons of material per year, the total cost of large scale geoengineering would be about one billion dollars a year. This sounds like a lot of money, but context is crucial: estimates put the worldwide monetary cost of climate change impacts in the neighborhood one trillion dollars per year by mid-century, even if we spend similar amounts to make deep emissions cuts. The cost of geoengineering the entire planet for a decade could be less than the $6 billion the Italian government is spending on dikes and movable barriers to protect a single city, Venice, from climate change-related sea level rise.

Today global spending on clean energy technologies is almost $300 billion per year—about a hundred times the direct cost of stratospheric aerosol geoengineering.

But I am not making a cost-benefit argument. I am not suggesting that one should choose geoengineering over clean energy. Whether or not we deploy geoengineering, we must still eventually decarbonize our energy system to reduce long-run climate risks along with the direct impacts of ocean acidification. I only claim that the direct cost of geoengineering is negligible compared to other climate related costs. Indeed, the low costs are disturbing because they could tempt some nation to leap into ill-considered trials or unilateral deployment. Costs are so low that they will not likely be a major determining factor in decisions about geoengineering. Economist Scott Barrett has called this "the incredible economics of geoengineering."

Limits

I have focused on a single method of solar geoengineering, stratospheric sulfate aerosol. With existing technologies and modest effort, we could begin using sulfates to alter the pace of climate change within a decade—probably faster than we can develop a broad-based scientific community with effective tools to monitor their efficacy and to look for unexpected problems, and almost certainly far faster than we can develop a stable international system for making decisions about their implementation. From Facebook to *in vitro* fertilization, technology moves ever faster than our ability to absorb it.

But climate is a century-scale problem.[35] What are the technical limits to solar geoengineering over this century? I will sketch a few advanced technologies that might emerge, not to predict what will happen, for that is a fool's errand, but rather to illustrate the potential for surprising outcomes. As more people begin to think about the social implications of geoengineering, they naturally tend to build their assumptions about the technology's risks and limits based on what is now known about sulfate aerosols or another technology, cloud whitening. Sulfate aerosols are (almost) where the story started; we would be foolish to expect that this is where it will end.

In guessing how quickly new technologies might emerge, one must be mindful that the slow pace of development since the pioneering work of Budyko and others in the 1970s is not an inevitable constant of geoengineering technology. Until a few years ago, essentially all work was an unfunded hobby by scientists with other research projects. If focused, large-scale research efforts begin, if geoengineering becomes a "trendy" topic, then we will see rapid innovation beyond sulfates. The dynamic of success-through-novelty is the mainspring of science and engineering.

The forcing/response distinction introduced in Chapter 3 neatly breaks the exploration of technical limits into two chunks: first, how precisely can we adjust radiative forcing of climate? And, second, given some level of control over radiative forcing how precisely might we alter the climate?

There are simple physical limits to the extent to which imposed solar radiative forcing can compensate for the risks of accumulating carbon; no manipulation of sunlight, for example, can stop the acidification of the oceans, which is a chemical effect of rising carbon dioxide concentrations.

There do not, in contrast, appear to be physical limits that prevent a technologically advanced civilization from tailoring radiative forcing in an arbitrary manner. There are—of course—limits to what we can do with current technology and a "reasonable" allocation of resources, but technology and resources are extraordinary labile over a century. Crude geoengineering—sulfates—is cheap enough that money is not an important constraint but more sophisticated, tailored, methods might get sufficiently expensive that cost would matter. To understand the technical limits it makes sense to consider methods that look expensive now. Today we shy away from spending even one percent of global GDP on cutting carbon emissions. We don't know how future generations will allocate their resources: they might well want to spend one percent of GDP rather than 0.01 percent on a geoengineering method that offered less risk and improved efficacy.

Edward Teller and Lowell Wood made the first systematic effort to explore the physical limits of geoengineering at Lawrence Livermore National Lab in the late 1990s. This fact stirred concerns about military involvement in geoengineering. While there is a long history of military interest in weather control, I am reasonably confident that there is no significant effort on geoengineering.[36]

Teller and Wood sketched a grab bag of physically plausible nano-structured particles that could scatter much more sunlight per unit mass than a droplet of sulfuric acid. These scattering designs offer advantages, such as the ability to tailor radiative forcing carefully and perhaps avoid some of the negative side effects of sulfate geoengineering. The trade-off is that they will likely be expensive to produce, and if they need to be renewed on an annual basis the cost could be prohibitive.

About a decade later, I began thinking about the physical limits of stratospheric scatterers and realized that one could use the photophoretic effect, a piece of early 20th century physics, to loft particles into the upper atmosphere and maintain them in a narrow altitude range. The essence of the idea is that if a particle is warmer than the surrounding air (as will generally be the case in the upper atmosphere), air molecules will bounce off the particle a little faster than they arrived.[37] The trick is that the probability of molecules of air "feeling" the warmth of the surface is not the same for all surfaces. If one made a little disk in which the lower side of the disc was made of a material for which air molecules were more likely to "feel" the warmth than the material of the upper side, then on average air molecules would bounce off the lower surface with larger velocity than they do off the upper surface. Newton taught us that action must equal reaction, so there will be a force pushing upwards from bottom to top. This is called the photophoretic force and can easily be strong enough to keep a small particle hovering in the upper atmosphere resisting the pull of gravity downwards. You might ask how the particle maintains its orientation? Why doesn't the disk flip upside down and feel the force driving downward? The answer is that you could use a material that orients itself in the Earth's electric field.

In effect it's a way to make flying saucers—discs that levitate with no apparent means of propulsion—but sadly the discs can't be much thicker than the size of a transistor in the CPU of your computer.

These little discs offer several potential advantages as tools to manipulate the earth radiative forcing. They might be engineered so they levitated above the stratosphere, thus removing the direct impact on stratospheric chemistry of the ozone layer. I say potential advantages because it is unclear whether they can be built and distributed at sufficiently low cost to be useful.[38]

Another advantage is that they reflect sunlight back to space without scattering much light downward towards the Earth. This gets around a problem with sulfate droplets or any other small roughly spherical scattering particle. All such particles scatter most of the light they intercept forwards, down towards the earth. In order to reduce the Earth's absorption of sunlight by one percent using sulfate aerosols one must scatter roughly nine percent of the light downwards, reducing the direct solar beam (the collimated light that casts sharp shadows) by ten percent. The net effect is akin to having a high layer of haze. This increase in diffuse light from sulfates or similar aerosols results in a host of subtle unintended effects, from increasing the productivity of most ecosystems and decreasing the effectiveness of concentrating solar power plants. The use of non-spherical scatterers such as the disks eliminates this problem.

One need not go to the stratosphere to alter the radiative forcing of climate. John Latham and colleagues have developed the idea of using very fine sea salt sprays that might be produced near the ocean surface. They would mix upward into low-lying marine clouds, making the clouds whiter and perhaps making them persist longer. At present the basic efficacy of this technique is far more uncertain than is stratospheric sulfate. No doubt that increasing the number of cloud condensation nuclei will, sometimes, make clouds whiter and more persistent. But the details are hideously complex. There are no models that can accurately relate the number of sea salt particles to the increase in large-scale cloud reflectivity under a realistic range of atmospheric conditions. In some circumstances, adding sea salt particles may paradoxically decrease reflectivity by breaking up the continuity of the cloud deck.[39] I can only conclude that there is a very strong case for research on sea salt spray.

It's all too easy for a tinkerer like me to get sucked into the technical details. And for the reader it is all too easy to get overwhelmed. The bottom line is simple. Because one can alter the entire climate with as little as 10,000 tons of super-efficient stratospheric scatters, an amount that could be lifted in a month by a single heavy lift stratospheric aircraft, there is extraordinary scope to develop new tools to allow more precise alteration of radiative forcing. None of these advanced technologies are ready for immediate implementation. It's impossible to know if they will be relevant over the next half century, but given the pace of technological progress it seems reasonable to suppose that if effort is expended to find "better" ways to alter radiative forcing, where better means more precise control in time and space and less environmental side effects, then such improvements will likely be found.

I am confident that we could eventually find "clean" ways to alter radiative forcing—methods that have negligible direct side effects. Precise alteration of radiative forcing does not, of course, enable one to precisely counteract climate change driven by greenhouse gases. "Eventually" entails stumbling around. Early attempts will no doubt have problems, and it will only be possible (though still not guaranteed) to converge on a good method by using a development pathway that has feedback loops that can catch bad ideas and fix them before they cause too much trouble.

Let me be clear how narrow my claim is lest I be accused of naïve techno-optimism. I am not claiming that we will necessarily develop such technologies. I am an optimist about the scope for technological innovation, but I am pessimistic (or, I think, realistic) about the social capacity to make wise use of new technology. Indeed, one can all too readily imagine horrific outcomes in which advanced geoengineering technologies could be used for destruction. If we develop the ability to cheaply and effectively manipulate radiative forcing, then in the theoretical worst-case these technologies can be used to create a worldwide Ice Age, a snowball earth such as existed about three quarters of a billion years ago in the so-called Neoproterozoic.[40] This is greater than the environmental disruption than could be achieved with the world's nuclear weapons. These technologies give humanity unprecedented leverage over global climate and that leverage can be used for good or ill. But if we do not develop them in a timely way, we may feel the unmitigated consequences of rapid warming, which could be a disaster for many hundreds of millions of people and the larger natural environment upon which we all depend.

How many knobs do you want on your climate control box?

Suppose that engineers could build anything physically possible into a system for controlling radiative forcing. What would minimize the chance of misuse, given the realities of human decision-making in a chaotic and unequal world?

The rapid response to changes in radiative forcing exemplifies the design choices at issue. Fast response means negligible delay between the decision to turn down radiative forcing and the physical decrease in net radiative forcing of the climate. Sulfate aerosols, for example, persist in the atmosphere for about a year, so a decision to stop adding sulfur will be felt a year later. Longer-lived particles like the levitating disks would respond more slowly. At the other extreme, consider a system of mirrors in space engineered so that each could be rotated on command to be either perpendicular to the Sun (blocking sunlight) or turned on edge (no effect on sunlight). The response time of such a system could conceivably be less than an hour.

Set aside technical feasibility for a moment and ask, what response time do we want? If the world were run by a single rational entity (think Spock), then one might as well have fast response because one can always choose to turn a fast-responding knob slowly, but without fast response you cannot adjust the knob quickly.

Still, I don't trust the world's current leadership with a fast control knob. The ability to turn off radiative forcing from geoengineering quickly is often called "the termination problem," but in fact it's both a blessing and a curse. All else equal, one wants a system that can respond quickly to new information. But the risk comes not from the rapid response time itself, but from its misuse. Political systems are prone to highly nonlinear responses. Too often nothing is done while a slow-moving problem slowly builds up steam; then, on occasion, precipitous decisions emerge from the maelstrom. Trouble could arise from two directions. On the one hand, politicians might overreact to indications that climate change was turning out to be worse than expected—picture dramatic crop failures in India and the U. S. following a particularly hot summer—turning on an untested system too quickly. On the other hand, if geoengineering was already in operation, politicians might overreact to some indication of dangerous side effects by turning the system off so quickly that the damages from the rapid change in radiative forcing were worse than the newly discovered side effects.

Similar trade-offs apply to the degree of regional control. If climate could be independently controlled in each region of the planet, then it might be fine have each region choose their climate without central coordination. But this is impossible. Regional climates are strongly interconnected by flows of wind and water that carry heat and moisture from place to place. It is physically possible to imagine a system that enables fast-acting local control of radiative forcing, but this does not enable independent control of local climate.

As with response time, a world of calmly rational entities that could negotiate trade-offs for mutual benefit might wish to have local control of radiative forcing. But, this might not be a wise choice in our world. Monsoons arise from the contrast between warm land and cool ocean. China already is concerned that its monsoon is weakening, threatening its ability to feed its population. Suppose China were to try to increase monsoon strength by cooling the Pacific Ocean off its coast by adjusting local radiative forcing, perhaps using the cloud brightening technique. The monsoons of Asia are strongly inter-connected by the large-scale flow of air around the planet.[41] It might be that an engineered increase in the Chinese monsoon will weaken the Indian monsoon, so the benefit to China would be a threat to India. How would we settle this? Both are nuclear weapons states. We lack even rough agreement about the norms of managing large-scale climate control, let alone an effective international mechanism for resolving disputes around it.

It might thus be better not to have local control of radiative forcing. This is a concern with the cloud brightening technique; it is inherently local and it only works, if at all, in limited locations with a specific kind of low-level marine clouds. In a world run by well-intentioned central planners, the ability to tailor radiative forcing using cloud brightening could be very helpful in maximizing the effectiveness of solar geoengineering while reducing harmful side effects. But in the world we live in, a technology enabling regional control might be more curse than blessing.

Returning to the theme of this chapter: developing tools for solar geoengineering is not a passive activity that discovers facts about nature, though some facts will no doubt be discovered. Rather, it is the deliberate application of existing knowledge to achieve specific goals. It is engineering, not science.

Sensible design requires public input

From the design of toasters to airplane cockpits, engineering programs now try to teach young engineers to think beyond narrow technical problems and to consider how users will interact with their designs. We need to build this ethic into development of geoengineering technologies. Public engagement is not a sideshow; it should be the central ring of this circus, for without public input the goals will be set implicitly by assumptions in the minds of the narrow technical community developing geoengineering.

As an active developer of these technologies, I know more about climate engineering than most but my values should count no more, or less, than those of anyone else. Knowledge alone does not tell us what we should do about climate change.

Scientists who say what *should* be done as if their recommendations sprang from an objective calculation are abusing public trust. Their value judgments have at least as large a role as facts in determining their recommendation.

Given the world as is, what goal should we give designers of solar geoengineering technology? Though not a goal, an entirely plausible answer is that we should wish there is no geoengineering technology available at all. My guess is many people wish that geoengineering would work if we really needed it in some "climate emergency," but that it should be somehow inaccessible, placed in a box with a "Break glass in case of emergency" sign to minimize the temptation to use it as a substitute for action to reduce emissions. In this line of thinking, the ideal outcome might be achieved if the technical community quietly developed the technology while the same time publicly exaggerated its risks.

Others may say we would be better off if geoengineering were buried. Underlying this view, I suspect, are some of the exaggerated claims about its risks, such as the threat it poses to the Asian Monsoon. But if some in the scientific elite truly hold this view, it is deeply arrogant. It assumes that they get to know truths that society is unequipped to handle and that they get to decide which truths are public and which not. Einstein gave the best response to this mind-set in a quote carved into a statue outside the U.S. National Academy of Sciences: "The right to search for truth implies also a duty; one must not conceal any part of what one has recognized to be true."

5

Ethics and Politics

I WAS EXPECTING FIREWORKS IN THE WOOD-PANELED confines of a Cambridge University hall as I helped introduce Jim Lovelock and Lowell Wood during the opening reception for a geoengineering meeting in 2004. In the late 1960s Lovelock proposed the Gaia hypothesis, the idea that earth can act like a single self-regulating organism. In 1971 he was the first to detect the accumulation of ozone destroying chlorofluorocarbons (CFCs) in the atmosphere, and later his popular books on Gaia established him as an icon of environmentalism. Wood is a brilliant and eclectic physicist, a former nuclear weapons designer and protégé of Edward Teller, and working with Teller, Wood had written a brilliant paper exploring the physics that form the outer boundaries of geoengineering. But no sparks flew as Drs. Gaia and Strangelove met.

One of the joys working this topic is its potential to create alliances where none exist. At the Cambridge event, Lovelock was eager to engage all aspects of geoengineering and told Wood of his regret at not being able to meet Teller.

It has also created divisions, especially among environmentalists. The most visible are sharply critical: filmmaker Sir David Attenborough equated it with fascism and Canadian environmentalist David Suzuki called it "insane." Quiet voices within major environmental groups urge cautious support for research on geoengineering. Mike Childs of Friends of the Earth and environmental writer Mark Lynas are among the few voices from within the environmental community to argue publicly for research. Perhaps the only formal initiative of any major environmental group has been the co-leadership of the Solar Radiation Management Governance by Environmental Defense.[42]

Critiques of geoengineering arise from diverse worldviews, and passions run very high. I have received two death threats that warranted calls to the police, and received many outraged comments from colleagues whom I respect. The most extreme critiques (and the death threats) have come from people who are convinced by the chemtrails conspiracy theory, which holds that the US government is deliberately spraying its citizens with toxins from aircraft. Believers claim that metals such as aluminum and barium are sprayed from commercial aircraft for purposes that are alleged to range from mass culling of the human population to mind control.[43] These views are widely held; one sixth of respondents in a large public survey we ran in Canada, Britain, and the United States believed that it was partially or completely true that "The government has a secret program that uses airplanes to put harmful chemicals into the air."[44]

Our poll found that people are more likely to oppose geoengineering if they are skeptical of government authority and self-identify with the right end of the political spectrum. While chemtrails believers are an extreme, they are part of a continuum that includes a much larger group which believes that climate risks are being exaggerated by the environmental left as an excuse to justify further extension of state power at the expense of individual freedoms. To overstate it, this view sees geoengineering as a tool used by a technocratic, transnational, and godless elite who have concocted both the climate threat and the geoengineering response as a means to extend their power.

On the left, the central complaint about geoengineering is that it presents a temporary and illusory technical fix, which encourages society to turn away from necessary social reforms. Corporate interests will exploit geoengineering for profit and as a means to defend the economic status quo on which they depend. Clive Hamilton, a left-leaning critic of geoengineering believes that "conservatives see it as a vindication of the system"—that market and technological innovation will solve our problems—

whereas "it is the political and economic system... that we have to change."[45] To overstate again, if the root cause of environmental ills is over-consumption driven by industrial capitalism then solutions must be fundamental social reforms, not new technologies that merely buy us more breathing space.

Not a moral hazard

The concern that geoengineering will encourage us to avoid hard choices about reducing carbon pollution is commonly and inaccurately called its moral hazard.

Moral hazard, as defined in economics, arises when individuals are shielded from the bad outcomes of a risky-but-otherwise-desirable activity so that they take on more risk than they would accept if they were paying the full cost. They get a free pass; but the cost ultimately falls on others. The presence of subsidized federal flood insurance, for example, encourages people to build (and rebuild) in flood zones because they do not pay insurance costs that reflect the true likelihood of devastating floods. Their welfare is being enhanced, but social welfare is not: a person who buys in the flood zone is responding sensibly to the fact that they get the benefits of living near the river while being shielded from some of the costs. But society is worse off because the true costs of living in a flood plain exceed the benefits derived from living close to the river.

Although I introduced the term into the geoengineering debate,[46] I don't think *moral hazard* does a good job capturing the concern geoengineering will weaken efforts to reduce emissions. With geoengineering, one might say that current generations get the free pass while the next generation pay the full cost, but this buries the tension between individual and social welfare and so wanders far from the original intent of *moral hazard*, and it obscures the likelihood that future generations will suffer more from our emissions than from geoengineering since the impact of carbon lasts far longer than any form of geoengineering now contemplated. Some use *moral hazard* to express concerns that powerful will get the pass while the poor pay the full cost. While concerns about inequality are and should be central to debate about geoengineering, it's misleading to equate inequality and moral hazard.

Risk compensation seems a better behavioral analog for fears that geoengineering will lead us deeper into the carbon-climate trap. Risk compensation is a change in behavior towards increased risk exposure after risk has been reduced by some technical fix. A defining example of risk compensation was the observation that driving fatalities decreased less than expected following the introduction of seatbelts, perhaps because belted drivers went slightly faster. A particularly striking example was a study that found drivers are 40 percent more likely to drive close to bicyclists who are wearing helmets than to those who are not.[47] Here the behavioral adjustment is not made by the cyclist who adopted the technical fix to reduce risk but rather by a second party, the driver, whose actions are imposing risk on the cyclist. This is apt because it's our generation who may emit more carbon and thus impose more climate risk on future generations because we anticipate that their risk will be reduced by future geoengineering (the generational equivalent of donning a bicycle helmet).

Risk compensation is not irrational. If one gets pleasure from risky sex or fast cars, and protective measures allow one to take a bit more pleasure with no additional risk, then risk compensation is a sensible response.

Most of us have some personal experience with this phenomenon. The existence of such wonders as satellite phones and Gore-Tex jackets enables me to do harder trips, to push farther into the wilderness than I would otherwise dare. I am more cautious—and more alert and alive—when I am soloing a winter trip in the mountains alone than when I have a satellite phone and know rescue is nearby.

The expectation that people will adjust their behavior does not provide an argument against introducing risk reducing technologies; it simply means that risk will be reduced less than one would predict using a calculation that ignored the behavioral change. It would be perverse in the extreme to argue that risk compensation justifies withholding condoms or seatbelts.

Similarly for geoengineering: all else equal, I expect a world where geoengineering is tested and available will be one that spends less on reducing emissions than a world where geoengineering was known to be impossible.

Frank discussion of risk compensation is rare. I have served on several high-profile committees that aimed to articulate formal consensus views about geoengineering. Each worked hard to reassure its audience that research on geoengineering should not in any way divert attention from cutting emissions. The Bipartisan Policy Center's report said "This task force strongly believes that climate remediation technologies are no substitute for controlling risk through climate mitigation"; and, "Message number one" of the Solar Radiation Management Governance Initiative was that "Nothing now known about SRM provides justification for reducing efforts to mitigate climate change through reduced GHG emissions, or efforts to adapt to its effects."[48]

I view these statements as (at best) confused. Why? They are no doubt motivated by a well-intentioned conviction that humanity is doing far too little to cut emissions in the face of the climate threat, a conviction I share unequivocally. But the fact that we ought to be cutting emissions so that we pass on less carbon-climate risk to our grandkids does not justify assertions that it would be wrong for humanity to alter its behavior if geoengineering does provide a meaningful reduction in climate risk.

The objective at the very root of mainstream climate policy is to improve human welfare by balancing the costs of reducing emissions against the damages from climate change. While I am personally unpersuaded by this utilitarian framing, it is a centerpiece of climate policy analysis.

If you accept the utilitarian view, then you are committed to accept the consequence that if climate risk is reduced then policy should reflect that change by slowing emissions cuts. The math is simple and unavoidable: less risk means less resources expended to insure against the risk. This is just as true were god to speak from the clouds (from where else would she speak on this topic?) announcing that she has reduced the climate's sensitivity to carbon dioxide, as it is if mere mortals were to achieve an imperfect and partial version of the same effect by injecting aerosols in the stratosphere.

If one wants to maintain that there should be no risk compensation—that we should work as hard as possible to cut carbon regardless of the potential of geoengineering—then one must adopt another framing for climate policy. It's tempting to argue that geoengineering is so uncertain that the mere possibility should not influence our behavior, but this argument is unsustainable. The analytical machinery of climate policy analysis is well equipped to cope with uncertainty in either the costs of cutting emissions or the risks of climate change.[49]

It is as silly to assert that geoengineering should only influence policy if it is certain to work as it is to claim that we should only cut emissions if climate impacts can be predicted with certainty.

Geoengineering that is only weakly effective still shifts utilitarian policy towards less mitigation just as insurers charge lower premiums if you install smoke detectors, notwithstanding the fact that they lower—but do not eliminate—the chance of a catastrophic fire.

This argument might seem to miss the point. There is a huge gap between actual progress being made to restrain emissions and the cuts prescribed by climate policy analysts. Fear that public discussion of geoengineering may make it harder to bridge that gap is, I suspect, the single most important basis for opposition to geoengineering research. Justified or not, this fear is distinct from the concern that—if the world followed a rational utilitarian policy—the feasibility of geoengineering would mean less mitigation.

There is no simple argument that shows geoengineering will expand the gap between *is* and *ought*.

Perhaps by making climate risks seem both more manageable and more dangerous geoengineering will shake up the stale politics of climate change and accelerate action. In any case, there is no basis for glib statements that geoengineering should not alter the amount that emissions ought to be cut.

Moral confusion about inequality

The heart of the risk compensation argument against geoengineering is that we should forgo the Band-Aid and tough it out while focusing on long-run solutions, because use of geoengineering would reduce the near-term climate impacts and so reduce our commitment to cutting emissions.

This argument becomes more ethically disturbing the more one thinks about the way risk and responsibility is divided between us and our grandkids, and between rich and poor.

It is odd even if one considers an imaginary, homogenous, democratic world without inequality. Recall how the slow dynamics of the carbon cycle separate emissions from their impacts. Put crudely, cutting emissions reduces the climate risk for our grandkids but it does nothing for us. Geoengineering does the opposite. It reduces climate risks for the generation that uses it—if it works at all—but does nothing to reduce the risks that our emissions impose on future generations.

We should reduce emissions now in order to reduce the climate risk we impose on the future. But this is not an argument against simultaneously using other means to reduce our near-term pain. If it was, then why not carry the argument to its absurd conclusion: If current pain is an effective spur to right behavior, then more pain should be better. Should we find ways to use fossil energy that impose extra harms on the current generation so that we are even more motivated to help our descendants? We might forego the use of proven technology to limit deadly particulate air pollution from coal-fired power plants even when the cost of pollution control is far less than the pure monetary cost of the death and disease they cause. As a deliberate policy this would be stupid and immoral (although in practice it is exactly what is happening in the U.S. and more egregiously in China[50]).

This is evident nonsense, but it gets worse. If the no Band-Aid argument is fallacious for a homogeneous world, it becomes truly repugnant when applied to the world we live in. Most emissions come from the rich. Most of the immediate near-term burden of climate change will fall on the poor and likewise it is the poor—because they are most vulnerable to climate change—who are most likely to benefit from any short-term reduction in climate risk offered by geoengineering.[51] Yet, most of those arguing that we should ignore geoengineering because it will divert us from "true" solutions—such as emissions cuts or deep rooted social change—are themselves wealthy and educated. So we have the ugly prospect of rich people arguing that we should reject the geoengineering Band-Aid—thus denying what may be a large benefit to the poor—in order to goad the rich into cutting emissions.

The developed-world advocacy organizations like the ETC Group—the most effective group opposing geoengineering research—claim to speak in the name of the poor and vilify those who advocate research on geoengineering.[52] They argue that the best solutions are a return to "peasant-led agricultural systems" in order "to guarantee global food security and survival of the planet."

The romantic embrace of the primitive is ever a tempting response to the powerlessness that many feel in the face of globalization. Yet the hard physical reality is that—despite the very real failures of modern agribusiness—the productivity of ancient peasant-led agriculture is far lower than is achieved with modern technological agriculture, so that if the world was to turn back the agricultural clock much of the world's population would soon die of starvation. Moreover, the carbon cycle's inertia dictates that even aggressive emissions will do little to address warming that is likely to damage crops by heat stress in the next quarter century, with resulting crop losses that will put millions of the poorest at risk.

This is moral confusion, not moral hazard. Emissions cuts and geoengineering are not interchangeable alternatives. We must cut emissions if we are to avoid passing the risks of our cheap energy to future generations. That is the hard and simple truth. Even if geoengineering can reduce risks in the future, it is still true that current emissions increase that risk, so geoengineering cannot absolve us from the duty to constrain our emissions.

That today's emissions pass a burden to the future does not in any way imply that we should forgo the possibility of using geoengineering to limit the harm that emissions by past generations have imposed on the present. This argument is only strengthened by the fact that emissions and energy use come mostly from the rich while the burdens of climate change fall most strongly on the poor who will also benefit most strongly from geoengineering (if it works) when it reduces these burdens.

Ends and means

Suppose it's true that a combination of geoengineering and emissions cuts could do a better job of limiting the environmental risks of climate change—such as extinction of cold adapted species by rapid Arctic warming—than could emissions cuts alone. Likewise, because the near-term human impacts of climate change—such as heat stress, flooding and crop failures—fall disproportionally on the world's poorest if it follows that the benefits of reducing climate change will go to people with the least ability to cope with environmental stress, that is, suppose the benefits of geoengineering may go primarily to the poor.

If geoengineering can protect the vulnerable and the natural environment, why then is there such strong opposition to geoengineering from the political left? The answer is at once obvious and obscure.

Since the 1960s the environmental left has evolved a set of widely held assumptions about the characteristics that mark good solutions to environmental problems. They include a preference for local over global, from community supported agriculture to the empowerment of indigenous peoples in the management of tropical forests; a preference for changing industrial processes to eliminate the pollutant over the use of end-of-pipe waste treatments; and finally, a preference for social over technological solutions. Geoengineering violates each of these assumptions and in that sense its rejection is unsurprising.

Activist Naomi Klein, for example, condemns geoengineering as a "rouge proposition."[53] In "Capitalism vs. the Climate" she argues that "real climate solutions are ones that steer these interventions to systematically disperse and devolve power and control to the community level, whether through community-controlled renewable energy, local organic agriculture or transit systems genuinely accountable to their users."[54] Maybe, but the link between these solutions and emissions reductions is tenuous, and Klein provides no evidence to make link other than asserting that big oil "got us into this mess." To pick just one example of fuzzy thinking, local organic agriculture does little or nothing to cut emissions, indeed it may have higher emissions than modern industrial farming.[55] It's hard not to suspect that the means and ends have been reversed, that Klein knows the political outcome she favors and has simply latched onto the climate threat as a way to advance it.

The tension between means and ends is stark: if the problem is that climate impacts fall unfairly on the poor and politically weak, why then is the solution emissions control and not specific measures to reduce poverty and empower the weak? It would be hard to argue that global emission control is an effective tool to fight poverty so I suspect the reason Klein favors emissions control is as a means to weaken the power rich rather than as an effective way to help the poor.

Climate change forces hard trade-offs. Are we trying to protect subsistence farmers threatened by drought or New Yorkers threatened by rising seas? Polar bears or wheat fields? It's tempting to answer, "All of the above." Tempting but glib. Cutting emissions will limit climate risk for future generations, yet it will increase current energy costs—with pain felt most strongly by today's poor. Trade-offs arise because there will be winners and losers as the climate changes, and also because cutting emissions is not the only way to limit the risks of a warming world. A sustained effort to cut emissions will—after many decades—reduce the risk of drought for vulnerable farmers, but the risks farmers face from changing climate can also be reduced by improving irrigation or by breeding crops that thrive in hot climates. A host of such local actions can help people adapt to a changing climate.

A deeper explanation for the rejection of geoengineering by many in the green left lies in the fact that environmentalists have conflated two different causes: repealing the excesses of aggressive capitalism and minimizing harm to the environment. I want action on both, but while they are connected, I don't see a tight one-to-one linkage between them.

Must we fix capitalism in order to fix the climate? Any serious argument in favor of this proposition must confront the fact that Western democracies have made enormous progress in managing environmental problems over the last half century. To cite just two examples, the abatement of air pollution by the regulatory system anchored by U.S. Clean Air Act has improved air quality even as population and wealth increased. This was no minor victory. These regulations imposed costs that peaked at almost 1% of the U.S. economy, they were opposed by powerful corporate interests and the political battle to enact them was long and fierce. The benefits in the form of improved health greatly exceed the cost and in many cities these regulations have improved life expectancies by almost a year. The fight is not over, but the victories are real and should be celebrated. A global restriction on ozone-destroying chlorofluorocarbons provides second and perhaps more impressive example because it demanded coordinated global action and it has succeeded in driving global emissions towards zero.

Clive Hamilton argues that North American scientists such as me bring "a distinctively American world view to geoengineering"[56] which he sets against more environmental views in Europe. While there is some truth here, it's worth recalling that the United States has led Europe in most major environmental legislation though that lead has (arguably) been reversed over the last two decades, and that it still leads in climate research and as a global center of non-governmental climate activism. While it's rhetorically convenient for Hamilton to equate the United States with corporate power, anti-environmentalism, and starry-eyed support for geoengineering, one might with an equal basis in fact, link America's leadership in environmental science and activism with its scientific community's interest in geoengineering.

The rise of inequality and the loss of social mobility in the United States are pressing examples of the need for political and economic reform, but to sustain claims that an effective response to climate change requires a fundamental reengineering of market capitalism is to deny the fact that liberal market economies have done a far better job of environmental regulation than their competitors.

It's time for environmental advocates to reexamine some old assumptions. Should we prefer local over global? It's an odd argument for the left to make. After all, the essence of solving global commons problems like climate is to compel local communities to cut their emissions to achieve a shared global good even though it's in the self-interest of each community to use that atmosphere as a carbon dump. The imperative that the global public good trump local autonomy is the reason environmental groups have expended such effort on global climate treaty. Arguments for local autonomy arguably come more naturally from the libertarian strain of the political right, a perspective that makes a paradox of the environmental left's preference for small-is-beautiful.

Should we prefer reengineering the production process so as to eliminate pollutants over end-of-pipe cleanup? Green Chemistry has made great progress using this approach. For me, it seems a wise rule of thumb and a sound basis for skepticism of technologies like carbon capture and geoengineering. But skepticism should not mean rigid opposition if these technologies offer real environmental benefits over their alternatives.

Finally, should we prefer social to technical fixes? One can argue either side, but whichever you prefer it's hard to avoid the conclusion that most of the big environmental wins of the last half century have been techno-fixes. Air pollution was cut using catalytic converters not carpooling; the ozone layer was saved by changing refrigerants not turning back to root cellars; and peregrine falcons were saved from DDT by developing insecticides less prone to bioaccumulation.

I am not advocating laissez-faire capitalism. On the contrary, I would like to see much more stringent environmental laws and a carbon price large enough to quickly drive high-carbon business out of existence. Getting to such an environmental victory will require a powerful new social movement. Indeed, looking backwards, one might turn my claim about historical environmental wins being technological on its head by arguing that the technology would never emerged without the rise of environmentalism and growing distrust of corporate power in the 1960s that drove the passage of the laws, laws that in turn drove industry to adopt the technical fixes that ultimately protected the environment. Klein rightly explains how corporate money spreads climate denial to block action.

The link between restraining aggressive capitalism and minimizing harm to the environment arises, in my view, not because political and economic liberalism are inherently anti-environmental, but because an accumulation of cooperate power and private political money has frustrated the ability of government to act in the public interest.

Risk, research, and lock-in

Even if you accept that concerns about moral hazard and inequality don't provide strong arguments against geoengineering, you may nevertheless feel that geoengineering sweeps something ugly under the rug.

I worry about our culture's tendency to overestimate the power of new technologies while underplaying their risks and about the human tendency to party today while putting off the clean up. These may be the strongest arguments for caution.

Perhaps the most common argument against geoengineering is simply that it will not work, or that risks will outweigh benefits. Though this is often folded into moralistic arguments, it is an empirical claim. And it may well prove true. If the question was: should we start geoengineering at full-scale today? Then, uncertainty would be a convincing argument against starting. But no sensible person advocates immediate commitment to large-scale geoengineering.

The question at hand is not immediate implementation, but whether or not to begin a serious research program that has strong mechanisms for independent review and is coupled with the development of governance mechanisms that could enable legitimate decisions about larger scale research and implementation.

One might argue that the chance that geoengineering will prove useful is so small that it's not worth research funds. It's true that we can't fund every wild idea; one must balance the chance of success against cost. But research seems particularly likely to reduce uncertainty in the case of geoengineering because so little serious work has been done, and relatively few of the well-developed tools that might improve understanding have yet been applied to the problem. Moreover, the cost is relatively small: the potential benefits in reduced climate damages are measured in hundreds of billions per year while the cost of a decade-long research program that would substantially improve understanding is measured in tens of millions.[57]

A much stronger argument against research is the potential for institutional lock-in that might make it impossible to forgo large-scale geoengineering once research develops momentum. This is a serious concern. Once established, commercial interests in geoengineering will no doubt defend the need to keep on keeping on, using whatever means available from backroom lobbying to funding science that tilts their way. The potential for lock-in is not confined to the private sector, so even elimination of private research cannot avoid the risk. Government institutions have an almost organic drive for self-preservation that leads them to exaggerate the original problem or create new missions long after their original mandate is obsolete. Academics defend positions literally to the death, though academia as a whole may be less susceptible to lock-in because individuals win large rewards by slaying sacred cows.

For me institutional lock-in is the only strong argument against low-risk research, but it's not a sufficient reason to abandon research. Rather, it's a spur to do all we can to counter such lock-in effects. We need to create new norms and develop rules such as restriction on patenting that exclude commercial interests from the core technologies. Competition and diversity are the best defense against lock-in. We should avoid dominance by single government research institutions and instead build a culturally diverse set of research and management efforts, with many explicitly devoted to figuring out all the ways that geoengineering will not work.

Inequality and geopolitics

Once developed, information technologies, human germ-line manipulation and geoengineering are relatively inexpensive to use and replicate. Each has the capacity to profoundly reshape the social order. Each has enormous leverage in that once the idea is known small inputs of materials and labor can yield huge output. In each case, it will be hard for nations or corporations that develop these technologies to maintain control over their dissemination and use.

These technologies differ profoundly in that countries or social groups can—within limits—make distinct choices about adoption. Some may forgo or control aspects of information technology while others embrace it. Some may choose to allow parents to manipulate their children's genes while others restrict the practice. Geoengineering is different. Local climates are so tightly coupled to their neighbors by flows of energy and moisture that local control is implausible. Geoengineering seems to demand centralized control. But whose hand will be on the thermostat?

Managing geoengineering demands global governance, and it will inevitably change the balance of power. Much rhetoric about the politics of geoengineering assumes that it will bolster the power of the strong over the weak, of the United States and Europe over the global south. While this is possible, the opposite seems just as likely. As a high-leverage technology for which intellectual property protection and secrecy are likely of little value, knowledge of how to implement geoengineering will spread globally, independent of where it originates.

Implementing geoengineering requires both knowledge and hardware. While the hardware may seem exotic, most of the technologies (high-altitude aircraft and simple dispersal technologies) could be produced by a surprising number of countries within a decade. All of the G20 countries have ample financial, technical, and institutional resources to develop and deploy the requisite hardware. The technology necessary to build an aircraft that could deliver payloads to the lower stratosphere is much less sophisticated than that necessary to build an advanced fighter aircraft. The job can be effectively accomplished with quite conventional jet aircraft technology.

Many less wealthy nations have ample financial and institutional resources to procure and control the hardware even if they could not easily make it themselves. Suppose, for example that the Alliance of Small Island States, a group that has played a vocal role in climate policy, decided to pool resources and begin work. They could choose between many credible suppliers from Hindustan Aeronautics Limited to Brazil's Embraer. The cost of developing the capability would amount to less than 1% of their GDP over a decade.

Any major military power could, of course, stop a small country from deploying such technology, either by threat of force or force itself. The credibility of that threat would depend, as always, on the extent to which the state or group of states who wished to begin geoengineering could recruit allies, perhaps by building a credible case that they were acting legitimately to protect their own interests without undue harm to nations outside the coalition.

In any case, the fact that these technologies could be readily afforded by most nations and that preventing deployment would require something at the level of embargo or military force is a sure sign that geoengineering is a leveling technology in the same ambiguous sense as the internet or nuclear weapons. If it comes down to a power struggle between nations over control of climate, geoengineering acts to diffuse power away from the richest and most powerful nations to a much larger pool of weaker states. Like other levelers—most notably nuclear proliferation—this fact is disturbing in its potential to lead to international conflict.

Geopolitics may be the hardest of all the challenges raised by geoengineering. When I consider geoengineering scenarios that lead to outright disaster, or converse scenarios in which geoengineering is prematurely abandoned despite its social and environmental benefits, all involve geopolitical failures.[58]

6

Prospect

Science and technology, like all original creations of the human spirit, are unpredictable. If we had a reliable way to label our toys good and bad, it would be easy to regulate technology wisely. But we can rarely see far enough ahead to know which road leads to damnation. Whoever concerns himself with big technology, either to push it forward or to stop it, is gambling in human lives.

–Freeman Dyson, *Disturbing the Universe* (1981)

THE SEAT PODS IN BUSINESS CLASS FACE INWARDS, away from the windows. Looking up from my work, I twist in my seat struggling to open the blind. Light floods in, washing out my neighbor's movie. Like spilling the "free" wine, it is a social blunder to let sunlight disturb the artificial world of a trans-Pacific flight.

At first I see only cloud cover. Then I am transfixed as I look straight down through a gap in the cloud deck onto a scalloped face of rock and snow that plunges to a bergschrund and then out to a glacier flowing between walls of snow-clad rock. Then, dimly, at the edge of perception a braided river appears through the thinning clouds. Views are made magical by the cloud's teasing, as with clothing that half-reveals the body underneath.

I want out. I want to walk the alder thickets along the braided river. I have been lucky to get far more than my share of time outside, but like an addict, wilderness just feeds my desire for more. Three years ago I skied for three weeks on the north coast of Baffin Island. Decades ago a friend and I threw ourselves in the mossy grass after hiking like madmen through the Talkeetna Mountains north of Anchorage in the same landscape that I now looking down upon. Drunk on the flinty joy of alpine scrambling at the limits of control, Curt and I ate berries as bears do, straining them off the bushes through our teeth. Wild country.

Now, I am flying to Korea for an IPCC meeting where we will speak about climate, nature, and geoengineering using only dry words in a dry room. I have spent late nights tweaking the Fortran of climate models when I might have been partying. I have seen hours and days flash past in sterile meeting rooms when I was trying to convince energy executives and politicians to get serious about cutting emissions when I might have been out in the woods with the rain on my face. Why?

My colleagues and I often argue that the money spent to cut emissions will be returned in reduced climate damages. This is not an incidental academic argument; rather, it is the canonical economic framing of climate policy, the root assumption of countless papers and policy briefs. According to this logic, my efforts to promote climate action have been a means to make the world a few percent richer in 2100 by saving money on climate impacts. I confess that I am not a believer. I find this utilitarian justification for climate policy utterly unconvincing. I do expect that emissions cuts will provide a net economic benefit, but I would still want to cut emissions even if doing so made us poorer.

In 1974 the constitutional scholar Laurence Tribe turned briefly to environmental matters in an article,

"Ways Not to Think about Plastic Trees," that was provoked by the installation of plastic trees in the median of a Los Angeles freeway. Tribe speculated that as we build artificial worlds our children will grow ever more contented within them so that interest in nature wanes. Values change with choices.

Our wealth now grows much faster than our carbon emissions, so it gets steadily easier to cut our carbon footprint. Over this century we could reduce emissions towards zero at the cost of a few years of economic growth. Yet, I fear that Tribe nailed it. A world in which the wealthy fly over magical country while facing inwards watching video screens in the dark is not a world that values nature.

I doubt that we can craft sensible plans for managing the planet while sitting in dry conference rooms.

The economist's story of optimal climate policy holds that we should make choices to maximize aggregate utility between ourselves and our (suitably discounted) grandchildren. Fair enough so far, but the story ignores the possibility that future values depend on current choices. Economists analyze how our choices influence our grandchildren's stocks of guns and butter, and yet the story assumes that our choices do not change how our grandchildren see the relative merits of guns and butter, or butter and glaciers. To the contrary, I suspect that choices we make about how to manage the natural world will influence the values of future generations, perhaps profoundly.

Because it is ostensibly value-neutral, the economic perspective provides a vital anchor for making policy in a multi-polar world. Paraphrasing Churchill, one might say that this utilitarian stance is the most unsatisfying basis for public policy except for all the others. I embrace this framework for making near-term policy trade-offs about air pollution or tax policy, but I think this analytical machinery is far less useful when pressed into service to make global-scale all-but-irreversible decisions that span centuries.

The economic framing that underpins climate policy is rooted in utilitarianism, the ethical stance that the right decisions are those that maximize collective welfare or "happiness." At risk of oversimplifying, a central problem with utilitarianism arises in balancing welfare between individuals. A standard trope of undergraduate philosophy is a scenario in which a group of sadists is deciding whether or not they should torture a lone individual. If the sadists get more collective pleasure (utility) from inflicting pain than the pain felt by the victim (disutility) then, under the most naïve interpretation of utilitarianism, torture would be the ethically correct choice.

The application to climate change is simple: if the majority get sufficient utility, or benefit, from cheap energy then it may be ethically right in a utilitarian sense for them to pollute if their benefits outweighs the costs to the victims of climate change.

The logic of economic utilitarianism explains—and justifies—social outcomes similar to the individual choices that yield behavioral risk compensation. A technology that (potentially) reduces environmental impact per unit of production may not have little effect on aggregate environmental footprint if its adoption increases consumption. The invention of the Haber process to produce artificial nitrogen fertilizers might have been used to intensify agricultural production and so decrease the total land footprint of agriculture, and it may have reduced the short term risk of starvation for some, and but the long-term effect was a further increase in global population and consumption. Human desire expands like a gas to fill the bottle defined by resources and technology.

If it works, geoengineering could be used to decrease humanity's environmental footprint; or, it could be used to increase what an ecologist might call the planet's carrying capacity for our species enabling an expansion in global consumption.

One way out of these utilitarian conclusions involves the attribution of inherent rights, moving ethics away from a simple balancing of welfare. Appealing to rights does not, however, resolve the ethical problems because one must face the question of what to do when rights conflict. How do we judge my "right" to enjoy the Arctic wilderness against your right to feed your family by mining coal? Do polar bears have rights? And, since it is central to the climate problem that current benefits are traded off against future risks, there are obvious questions about how to think about the rights of future generations.

I suspect that attitudes toward geoengineering and climate change turn on our personal attitudes about technology and nature more than is often acknowledged. If we consider only strictly utilitarian reasons for climate action then it is difficult to justify emissions cuts of the magnitude that many—including me—would like to see. Likewise, assuming geoengineering works as well as the early science suggests it does, then it is hard to avoid the conclusion that some geoengineering should be implemented.

Professionals who argue for action on climate change accept as a matter of commonplace wisdom that old-fashioned environmental arguments for preserving the natural world have little power to move the public, that it's more effective to make climate change seem real and threatening by tying it to issues of immediate public concern. The (supposedly) more effective arguments portray climate as a security problem, a food problem, a water problem or even an opportunity for jobs creation and economic competitiveness. These arguments are not without merit. But my hunch is that mobilizing large-scale action will require us to talk directly about climate as a new kind of planetary environmental crisis, not because it poses a dire threat to our welfare, but a crisis nevertheless in the future of a natural world squeezed by industrial civilization.

The current approach to environmental advocacy turns on the assumption that people will only respond to arguments from self-interest. The reason that commitment to cut emissions have been feeble, the argument goes, is the simple fact that it's hard to motivate people to take costly action when most of the benefits of that action are distributed globally and fall more than half a century in the future. Self-interest is a powerful motivator, yet selfishness and short-term thinking are an incomplete explanation for inaction, for these two tendencies are always at play and yet we have made progress on other challenges that required collective investment and long-term thinking. Turning up the climate-catastrophe hype has not been effective to date—disaster fatigue sets in fast—so perhaps it's time for something new: let's speak forthrightly about the non-utilitarian reasons to preserve a thriving natural world.

From fire-hunting to genetically engineered crops, our species has spent millennia using technology to alter the environment. Geoengineering is most often seen as a way to limit the impact of our carbon emissions on climate, but even if the technology is first used in the service of limiting environmental impacts I doubt its use will end there. For instance, if I were hired to maximize global agricultural productivity I might engineer a climate different than today's, one with higher carbon dioxide concentration to encourage plant growth, in combination with geoengineering to limit excessive temperatures. Geoengineering gives humanity new powers to shape the planetary environment; history suggests it's unlikely that this power will be used only to limit the environmental impacts of other human actions.

Economist and Nobel Laureate Tom Schelling identified the mutability of geoengineering's objectives in 1982. In a United States National Academy of Sciences study of climate change, he wrote that, "Interest in CO_2 may generate or reinforce a lasting interest in national or international means of climate and weather modification [*geoengineering*]; once generated, that interest may flourish independent of whatever is done about CO_2."[59]

If you want to leave your great grandkids with a vibrant civilization *and* to protect more of the natural heritage of the planet then you can justify strong actions to limit climate change, including geoengineering. I believe this despite the fact that by altering all of the world's landscapes carbon-driven climate change represents the "end of nature" as Bill McKibben argued in his 1989 book of the same name.

With or without geoengineering it's the end of nature with a capital 'N,' the romantic ideal of nature wholly separate from civilization. It's my hope that deliberate management of climate change—including geoengineering—can be the beginning of a renewed commitment to build a thriving civilization that honors its intimate connection to the natural world.

At its root, a decision to cut emissions, or to geoengineer, or both, or neither, demands an extension of our moral compass to include beings distant from our day-to-day world: future generations, the distant poor, and, the natural world. No basket of technical fixes will solve the carbon-climate problem if humanity cannot reach some rough social consensus about shared values that drive action.

Should we include deliberate planetary-scale engineering as part of our climate toolbox? If so, how should we choose the mixture of emissions cuts, adaptation and geoengineering that are used to manage climate risks? I think we should develop geoengineering and use it soon if it can safely slow climate change, but no one can extract objective answers to these questions from science. The answers turn not simply on facts, but equally on the overarching goals of action to limit climate change, goals that in turn depend on values—not science.

Geoengineering often seems a joyless choice between unpleasant alternatives. As journalist Eli Kintisch put it, "A bad idea whose time has come." But I can't wholly embrace that view. It's an easy out.

About a million years after inventing stone cutting tools, ten thousand years after agriculture, and a century after the Wright Brothers flight, humanity's instinct for collaborative tool building has brought us the ability to manipulate our own genome and our planet's climate. These tools rest on deep knowledge of the natural world accumulated over centuries. This knowledge was built by the efforts of countless individuals—all filled with error and motivated by self-interest—yet each also contributing to the accumulation of understanding. We may use these powers for good or ill, but it is hard not to delight in these newfound tools as an expression of collaborative human effort to understand the natural world.

The path towards wise use of geoengineering may require us to simultaneously grasp the delight in our new tools along with the humility to see their limits.

Acknowledgments

IDEAS ARE AN INHERENTLY SOCIAL PRODUCT. The ideas I have knit together into this book have arisen from academic collaborations and countless conversations with friends, critics, and colleagues. I am particularly indebted to collaborators and mentors including Jim Anderson, Jay Apt, Ken Caldeira, Bill Clark, Hadi Dowlatabadi, John Dykema, Steve Hamburg, Tad Homer-Dixon, Jane Long, Doug MacMartin, Ashley Mercer, Juan Moreno-Cruz, Granger Morgan, Andy Parker, Ted Parson, Kate Ricke, John Shepherd, David Victor, and Debra Weisenstein. It's a joy to work on a topic which generates strong opinion in almost everyone, and my thinking has been shaped by many chance conversations with concerned citizens from cab drivers to schoolchildren.

The text and its errors are my responsibility. As an academic author new to writing for a general audience I have benefited hugely from professional editing by Deb Chasman and Hal Clifford and from a few careful and generous readers who have commented on drafts including Jody Blackwell, Tony Keith, Oliver Morton, Andy Parker, and Daniel Thorp.

NOTES

[1] Cheryl Jones. "Climate not to blame for the extinction of Australia's big animals." *Nature*, 23 January 2010 www.nature.com/news/2010/100123/full/news.2010.30.html; and, Gilford H. Miller, et al. "Pleistocene Extinction of *Genyornis newtoni*: Human Impact on Australian Megafauna." Science, 283: 205-208 (1999). doi:10.1126/science.283.5399.205.

[2] I made this argument in more depth in congressional testimony, see: http://science.house.gov/sites/republicans.science.house.gov/files/documents/hearings/020410_Keith.pdf

[3] I try to keep a wall between Carbon Engineering (www.carbonengineering.com) and my academic work. I do not get involved in academic assessments of air capture nor do I talk about it in public unless it I am clearly identified as a representative of industry rather than as an academic, and in such events, I focus on Carbon Engineering's work and avoid offering general "academic" comments on technology or public policy.

[4] These numbers are rounded for simplicity. The amount of sulfur needed to offset a given amount of radiative forcing has an uncertainty of at least 30%. Note also, that the rate of

growth of greenhouse gasses and their associated radiative forcing depends—of course—on future emissions.

[5] These specific ideas originated with a study by Aurora Flight Science Corporation that I helped to initiate and manage. See: Justin McClellan, David W. Keith, and Jay Apt. "Cost analysis of stratospheric albedo modification delivery systems." *Environmental Research Letters*, 7:034019 (2012). doi:10.1088/1748-9326/7/3/034019.

[6] Juan Moreno-Cruz et al.. "A simple model to account for regional inequalities in the effectiveness of solar radiation management." *Climatic Change*, 110:649-66 (2012). doi: 10.1007/s10584-011-0103-z. Available at: www.keith.seas.harvard.edu/geo.html. Katharine Ricke et al. "Regional climate response to solar-radiation management." *Nature Geoscience*, 3:537-541 (2010). doi:10.1038/ngeo915. Douglas G. MacMartin et al. "Managing tradeoffs in geoengineering through optimal choice of non-uniform radiative forcing." *Nature Climate Change*, (2012). doi: 10.1038/NCLIMATE1722.

[7] "Once you get to the point in terms of climate changes that you feel you have to use it, then you have to use [SRM] forever," Pierrehumbert as quoted in: David Rotman. "A Cheap and Easy Plan to Stop Global Warming." *Technology Review*. February 2013. www.technologyreview.com/featuredstory/511016/a-cheap-and-easy-plan-to-stop-global-warming/

[8] Jane Mayer. "Taking it to the streets." *The New Yorker*, 28 November 2011. www.newyorker.com/talk/comment/2011/11/28/111128taco_talk_mayer.

[9] Humanity controls its industrial emissions, but the concentration of carbon dioxide depends not just on our emissions but also on how carbon is exchanged with soils, forests and the ocean. Each of these carbon flows is influenced by human activities other than fossil fuel combustion each as its own dynamics and uncertainties. Overall, this uncertainty in the carbon cycle is a relatively small contributor to the total uncertainty in linking emissions to climate risk, but all uncertainties multiply up.

[10] *The Economist Intelligence Unit.* "Rock steady: A special report on coal demand." The Economist Intelligence Unit, May 2013. www.eiuresources.com/Coaldemand2013.

[11] An increase in carbon dioxide warms the climate and that, in turn, alters the flow of carbon in or out of the atmosphere via processes as diverse as the rate at which organic matter decays (warmer wood rots faster) and the solubility of carbon dioxide in ocean water (warmer water holds less carbon dioxide). In the full interacting system one cannot unambiguously separate cause from effect or forcing from response. When we speak of climate forcing by carbon dioxide we are using an analytical trick, albeit one that is crucial to understanding climate change. We are defining climate's response to carbon dioxide forcing as the change that would happen if we could control the amount of carbon dioxide in the air while allowing everything else to run freely. Such experiments are, of course, easily performed on computer models, and our experience with those models in turn shapes the language and thinking of atmospheric science.

[12] Ken is a wunderkind and a bit of a bad-boy of climate science, a former anti-nuclear activist who helped organized a huge 1982 New York rally against Reagan's arms buildup that

I attended fresh out of high school. Ken and I now work together on geoengineering, and all three of us serve as advisers to Bill Gates on energy and environmental issues.

[13] At that time both Ken Caldeira and Lowell Wood worked at Lawrence Livermore National Laboratory. Ken's first paper on geoengineering is: Bala Govindasamy and Ken Caldeira. "Geoengineering Earth's radiation balance to mitigate CO2-induced climate change." *Geophysical Review Letters*, 27: 2141-2144 (2000). doi:10.1029/1999GL006086.

[14] We first add up the precipitation error for all regions weighting the regions by population so that a region with twice the population count twice as much in the average; then, we adjust the amount of solar geoengineering to minimize this total error. See footnote 6.

[15] Alan Robock, Luke Oman and Georgiy L. Stenchikov. "Regional climate responses to geoengineering with tropical and Arctic SO2 injections." *Journal of Geophysical Research: Atmospheres*, 113:D162008 (2008). doi:10.1029/2008JD010050.

[16] As of 11 April 2012 a Google search for *geoengineering* and *monsoon* yields 0.88 million hits while a search for *geoengineering* alone yields 1.56 million.

[17] Arun Gupta. "Geoengineering the Planet? Corporate remaking of the Earth's atmosphere." *Z Magazine*, 23.6 (2010) <http://www.zcommunications.org/geoengineering-the-planet-by-arun-gupta>.

[18] *Climate Change 2007: Working Group II: Impacts, Adaptation and Vulnerability*. Cambridge University Press, Cambridge,

United Kingdom. See section 19.3.6 "Extreme events" at www.ipcc.ch/publications_and_data/ar4/wg2/en/ch19s19-3-6.html.

[19] See figure S5 in: J. Pongratz, D. B. Lobell, L. Cao and K. Caldeira. "Crop yields in a geoengineered climate." *Nature Climate Change*. 2:101–105 (2012). doi:10.1038/nclimate1373.

[20] Alan Robock et al. "A Test for Geoengineering?" Science. 327:530-531 (2010). doi:10.1126/science.1186237

[21] John Shepherd, et al. *Geoengineering the climate–Science, governance and uncertainty*. The Royal Society, 2009 http://royalsociety.org/uploadedFiles/Royal_Society_Content/policy/publications/2009/8693.pdf.

[22] It's a case of life protecting life, the development of land-based life was not possible until photosynthetic organisms made sufficient oxygen to produce an ozone layer.

[23] An ozone molecule has three oxygen atoms. It is formed by adding one "odd" oxygen to the pair for atoms that forms a regular oxygen molecule. A central lesson of environmental science over the last half-century is that if one invents a new chemical one must understand the full path of its breakdown products in the environment. If chemicals don't break down quickly in the environment they may accumulate, reaching dangerous levels in unexpected places, as was the case for pesticide residues accumulating in fatty tissues.

[24] Simone Tilmes, et al. "Impact of geoengineered aerosols on the troposphere and stratosphere." *Journal of Geophysical Research*, 114:D12305 (2009). doi:10.1029/2008JD011420.

[25] The World Health Organization estimates that outdoor air pollution causes 1.3 million deaths per year worldwide. See: www.who.int/mediacentre/factsheets/fs313/en/index.html

[26] For a thoughtful and succinct view of how technology differs from science see: W. Brian Arthur. *The Nature of Technology: What It Is and How It Evolves*. London: Penguin Group, 2009.

[27] The "ozone hole" is the very rapid depletion of ozone in the North and South polar stratosphere that arises from circumstances particular to the polar stratosphere (removal of nitrogen oxides), in addition to the Molina and Rowland mechanism.

[28] Douglas MacMynowski et al. "Can we test geoengineering?" Energy and Environmental Science, 4: 5044-5052 (2011). doi: 10.1039/c1ee01256h

[29] Wallace S. Broecker. "Unpleasant Surprises in the Greenhouse?" *Nature*, 328: 123-126 (1987). doi:10.1038/328123a0.

[30] Jeffrey R. Pierce, et al. "Efficient formation of stratospheric aerosol for geoengineering by emission of condensible vapor from aircraft." *Geophysical Research Letters*, 37:L18805 (2010). doi:10.1029/2010GL043975. Available at: www.keith.seas.harvard.edu/geo.html.

[31] Crutzen, P.J. "Albedo Enhancement by Stratospheric Sulfur Injections: A Contribution to Resolve a Policy Dilemma?" *Climatic Change*, 77:211-219 (2006). doi:10.1007/s10584-006-9101-y.

[32] If the lifetime is ten times shorter it will takes ten times as much aerosol to produce the same radiative forcing, and so costs and some side-effects such as air pollution will be tenfold worse.

[33] These estimates are based on a consulting study by an aerospace consulting company with experience developing high-altitude aircraft (see McClellan et al. cited in endnote 5). Twenty Gulfstream C-37A's allows about 0.45 million tons per year to the stratosphere. Assuming sulfur payload with conversion to SO_3 in flight and using the results of Pierce et al. (cited in endnote 26) this gives about 0.4 Wm^{-2}.

[34] The direct vapor injection scheme (endnote 26) we proposed as a partial fix for the big droplet problem would not likely be possible using a reasonable number of hoses, so one can argue that the hose system would likely not work well for large-scale sulfate aerosol geoengineering.

[35] Dramatic changes in climate or deployed technology take decades. While the climatic impacts of our emissions extend millennia into the future, navigating the carbon-climate challenge is a job for this century.

[36] E. Teller, L. Wood, and R. Hyde. *Global Warming and Ice Ages: I. Prospects for Physics Based Modulation of Global Change.* Lawrence Livermore National Laboratory, 1997. Available at: http://dge.stanford.edu/labs/caldeiralab/Caldeira%20downloads/Teller_etal_LLNL231636_1997.pdf. For a summary of weather control and military involvement linked to geoengineering see: David W. Keith. "Geoengineering the Climate: History and Prospect." *Annual Review of Energy and the Environment*, 25: 245-284 (2000). doi:10.1146/annurev.energy.25.1.245. Available at: www.keith.seas.harvard.edu/geo.html.

[37] Molecules in a hot gas move faster than when the gas is cold, this is the essence heat is at the molecular level.

[38] Simple nanofabricated materials can be made at low cost, but structures I propose are more complex and I do not know if they can be made at sufficiently low cost, see: David W. Keith. "Photophoretic levitation of engineered aerosols for geoengineering." *Proceedings of the National Academy of Sciences,* 107: 16428-16431 (2010). doi:10.1073/pnas.1009519107.

[39] H. Wang, P. J. Rasch, and G. Feingold. "Manipulating marine stratocumulus cloud amount and albedo: a process-modelling study of aerosol-cloud-precipitation interactions in response to injection of cloud condensation nuclei." *Atmospheric Chemistry and Physics,* 11:4237-4249 (2011). doi:10.5194/acp-11-4237-2011.

[40] A snowball earth state might be triggered by 8% cut in sunlight over a century, this is about four times the reduction in sunlight that is often considered for geoengineering. See J. Yang, W. R. Peltier and Y. Hu. "The initiation of modern soft and hard Snowball Earth climates in CCSM4." *Climate of the Past.* 8:1-29 (2012). doi:10.5194/cpd-8-1-2012.

[41] Picture the way the Jetstream wiggles up and down across the mid-latitudes, if you push the Jetstream northward in one place it will wiggle southward somewhere else.

[42] www.srmgi.org/about-srmgi

[43] Evidence offered to support this view is often little more than the conviction that aircraft contrails persist longer than is natural. Wikipedia provides a good overview of the conspiracy theory, http://en.wikipedia.org/wiki/Chemtrail_conspiracy_theory. There are many good websites devoted to debunking chemtrails

including http://conspiracies.skepticproject.com/articles/chemtrails/ and http://contrailscience.com.

[44] The survey was run by my PhD student Ashley Mercer and administered by survey firms that guarantee representative samples. We surveyed about 3000 people in Canada, Britain and the US in 2010. See A. M. Mercer, D. W. Keith, and J. D. Sharp. "Public understanding of Solar Radiation Management", *Environmental Research Letters*, 6:044006 (2011). doi:10.1088/1748-9326/6/4/044006.

[45] Clive Hamilton interviewed by Amy Goodman of Democracy Now, 20 May 2013, see: http://m.democracynow.org/stories/13653 . The full quote is "They see it as a—the conservatives see it as a vindication of the system. They see it—see geoengineering as a way of protecting the system, of preserving the political economic system, whereas others say the probably is the political and economic system, and it's that which we have to change."

[46] To my knowledge the first use of *moral hazard* in conjunction with geoengineering was in 2000 in my article "Geoengineering the Climate: History and Prospect," cited in endnote 32.

[47] Ian Walker. "Drivers overtaking bicyclists: Objective data on the effects of riding position, helmet use, vehicle type and apparent gender." *Accident Analysis & Prevention*, 39: 417–425 (2007). doi:10.1016/j.aap.2006.08.010. The Wikipedia entry on Risk Compensation serves as a good introduction to the overall topic, and a classic early paper is Gerald J. S. Wilde. "The Theory of Risk Homeostasis: Implications for Safety and Health." *Risk Analysis*, 2: 209–225 (1982). doi:10.1111/j.1539-6924.1982.tb01384.x.

[48] Bipartisan Policy Center. *Task Force on Climate Remediation Research.* Washington DC, USA, 2011 http://bipartisanpolicy.org/library/report/task-force-climate-remediation-research; Solar Radiation Management Governance Initiative, *Solar radiation management: the governance of research*, Buckinghamshire, UK, 2011. www.srmgi.org/report.

[49] Juan B. Moreno-Cruz and David W. Keith. "Climate Policy under Uncertainty: A Case for Geoengineering," *Climatic Change*, (2012). doi:10.1007/s10584-012-0487-4. We added geoengineering to an entirely standard treatment of uncertainty in climate policy that balances current costs (emissions cuts) against future benefits (less climate impacts) in the face of uncertainty about both climate impacts and geoengineering. The specific numerical results depend on our assumptions (guesses) but the direction of the conclusion is strong: because it serves to limit the worst-case outcomes, geoengineering can play a valuable role in limiting the overall cost of climate change even when we assume it has significant risks and limited effectiveness.

[50] The most important source is the congressionally mandated EPA study, U.S. Environmental Protection Agency (1997), "The Benefits and Costs of the Clean Air Act, 1970 to 1990" available at http://www.epa.gov/cleanairactbenefits/. The study found economic benefits such as reduced hospitalization were roughly ten times larger than the costs of cutting emissions. Despite this many cost-effective air pollution technologies are have not been implemented in the U.S. and around the world, a gross failure of public policy. One has to wonder how we can expect to make progress on climate change when we can't manage air pollution.

[51] Poor and rich are now, of course, intermingled as the rich include the hundreds of millions who have rich-world living standards and carbon emissions in poor countries like China, while at the same time there are poor communities in the rich world that are tied to marginal fishing or farming that may bear significant climate burdens.

[52] The ETC Group's mission statement reads "ETC Group works to address the socioeconomic and ecological issues surrounding new technologies that could have an impact on the world's poorest and most vulnerable people." See http://www.etcgroup.org. The ETC group argues that geoengineering is profiteering. Perhaps the strongest single quote on the ETC websites is from John Vidal of The Guardian, who said of the authors of the Bipartisan Policy Center report (cited in endnote 44) "this coalition of US expertise is a group of people which smell vast potential future profits for their institutions and companies in geo-engineering."

[53] Naomi Klein, "Geoengineering: Testing the Waters", *New York Times*, 27 October 2012. www.nytimes.com/2012/10/28/opinion/sunday/geoengineering-testing-the-waters.html?.

[54] Naomi Klein, "Capitalism vs. the Climate", *The Nation*, 28 November 2011. www.thenation.com/article/164497/capitalism-vs-climate.

[55] Christopher L. Weber and H. Scott Matthews. "Food-Miles and the Relative Climate Impacts of Food Choices in the United States," *Environmental Science and Technology*, 42:3508–3513 (2008). doi:10.1021/es702969f.

[56] Clive Hamilton. "The clique that is trying to frame the global geoengineering debate," *The Guardian*, 5 December 2011. www.guardian.co.uk/environment/2011/dec/05/clique-geoengineering-debate.

[57] Moreno-Cruz and Keith cited in endnote 45; and, Ken Caldeira and David W. Keith. "The Need for Climate Engineering Research." *Issues in Science and Technology*. 27: 57-62 (2010). www.issues.org/27.1/caldeira.html.

[58] David W. Keith and Andy Parker. "The fate of an engineered planet." *Scientific American*, 308:34-36 (2013). doi:10.1038/scientificamerican0113-34.

[59] Board on Atmospheric Sciences and Climate, Commission on Physical Sciences, Mathematics, and Resources, National Research Council "Changing Climate: Report of the Carbon Dioxide Assessment Committee," National Academy Press, Washington, DC (1983).

ABOUT THE AUTHOR

David Keith is Professor of Public Policy at the Kennedy School and Gordon McKay Professor of Applied Physics in the School of Engineering and Applied Sciences at Harvard University. He has worked near the interface between climate science, energy technology and public policy for twenty years. He took first prize in Canada's national physics prize exam, won MIT's prize for excellence in experimental physics, and was listed as one of *TIME* magazine's Heroes of the Environment 2009.

BOSTON REVIEW BOOKS

Boston Review Books is an imprint of *Boston Review*, a bimonthly magazine of ideas. The book series, like the magazine, covers a lot of ground. But a few premises tie it all together: that democracy depends on public discussion; that sometimes understanding means going deep; that vast inequalities are unjust; and that human imagination breaks free from neat political categories. Visit bostonreview.net for more information.

The End of the Wild STEPHEN M. MEYER

God and the Welfare State LEW DALY

Making Aid Work ABHIJIT VINAYAK BANERJEE

The Story of Cruel and Unusual COLIN DAYAN

Movies and the Moral Adventure of Life ALAN A. STONE

The Road to Democracy in Iran AKBAR GANJI

Why Nuclear Disarmament Matters HANS BLIX

Race, Incarceration, and American Values GLENN C. LOURY

The Men in My Life VIVIAN GORNICK

Africa's Turn? EDWARD MIGUEL

Inventing American History WILLIAM HOGELAND

After America's Midlife Crisis MICHAEL GECAN

Why We Cooperate MICHAEL TOMASELLO

Taking Economics Seriously DEAN BAKER

Rule of Law, Misrule of Men ELAINE SCARRY

Immigrants and the Right to Stay JOSEPH H. CARENS

Preparing for Climate Change MICHAEL D. MASTRANDREA & STEPHEN H. SCHNEIDER

Government's Place in the Market ELIOT SPITZER

Border Wars TOM BARRY

Blaming Islam JOHN R. BOWEN

Back to Full Employment ROBERT POLLIN

Occupy the Future DAVID B. GRUSKY, ET AL.

Giving Kids a Fair Chance JAMES HECKMAN

The Syria Dilemma NADER HASHEMI & DANNY POSTEL